Practical Numerical Mathematics with MATLAB

Solutions

Other World Scientific Titles by the Author

Numerical Linear Algebra
ISBN: 978-981-122-389-1
ISBN: 978-981-122-484-3 (pbk)

Practical Numerical Mathematics with MATLAB
Solutions

Myron M Sussman
University of Pittsburgh, USA

World Scientific

NEW JERSEY · LONDON · SINGAPORE · BEIJING · SHANGHAI · HONG KONG · TAIPEI · CHENNAI · TOKYO

Published by

World Scientific Publishing Co. Pte. Ltd.

5 Toh Tuck Link, Singapore 596224

USA office: 27 Warren Street, Suite 401-402, Hackensack, NJ 07601

UK office: 57 Shelton Street, Covent Garden, London WC2H 9HE

British Library Cataloguing-in-Publication Data
A catalogue record for this book is available from the British Library.

PRACTICAL NUMERICAL MATHEMATICS WITH MATLAB
Vol. 1: A Workbook/Vol. 2: Solutions

ISBN 978-981-124-600-5 (set_hardcover)
ISBN 978-981-124-518-3 (set_paperback)

ISBN 978-981-124-035-5 (vol. 1_hardcover)
ISBN 978-981-124-519-0 (vol. 1_paperback)
ISBN 978-981-123-918-2 (vol. 1_ebook for institutions)
ISBN 978-981-124-070-6 (vol. 1_ebook for individuals)

ISBN 978-981-124-069-0 (vol. 2_hardcover)
ISBN 978-981-124-520-6 (vol. 2_paperback)
ISBN 978-981-124-433-9 (vol. 2_ebook for institutions)
ISBN 978-981-124-434-6 (vol. 2_ebook for individuals)

For any available supplementary material, please visit
https://www.worldscientific.com/worldscibooks/10.1142/12340#t=suppl

Desk Editor: Yumeng Liu

Printed in Singapore

Preface

This book contains solutions to the exercises in the workbook, "Practical Numerical Mathematics with MATLAB : A Workbook." All of the exercises in that workbook are included here, generally in the format described in the templates that accompany the workbook. The code included here follows the same guidelines given in the workbook. Exercise numbering in the solutions agrees with the numbering in the workbook.

Contents

PART 1

Rootfinding, Interpolation, Approximation and Quadrature

Chapter 1

Introduction to MATLAB

Solution 1.1.

(1) pi = 3.1416

  eps = 2.2204e-16

  realmax = 1.7977e+308

  realmin = 2.2251e-308

(2) In long format, pi = 3.14159265358979

(3) pi - 3.1416 = -7.3464e-06, Is it zero? (<u>yes</u>/no)

  (Long format pi)-pi = a roundoff-sized number

(4) What is the difference in the way that MATLAB displays a and b? b has trailing zeros

  Can you tell from the form of the printed value that a and b are different? (<u>yes</u>/no)

(5) Will the command **format long** cause all the decimal places in b to be printed, or is there still some missing precision? (<u>yes</u>/no) b still has all trailing zeros, no final nonzero digit

(6) Are the values of c and d different? (yes/<u>no</u>)

(7) x = student's preference

(8) Square of x = x^2

  Cube of x = x^3

(9) theta = student's preference

(10) $\sin\theta$ = correct value

  $\cos\theta$ = correct value

  Degrees or <u>radians</u>?

(11) In your own words, what is the difference between the following two expressions?

  a1='sqrt(4)'

  a2=sqrt(4)

  a1 is a string, a2 is the number 2

(12) What is the result of the command eval(a1)? 2

  Of a3=6*eval(a1)? 12

(13) Has the file `myfile.mat` been saved? (yes/no)
(14) After `clear`, is the "current workspace" empty? (yes/no)
(15) After `load myfile.mat`, has the "Current workspace" been restored? (yes/no)

Solution 1.2.

(1) What is the command you used? `meshPoints=linspace(-1,1,1000);`
(2) What expression will yield the value of the 95th element of `meshPoints`?
`meshPoints(95)`
hat is this value? -0.81181
(3) Does the variable `meshPoints` appear in the "Current workspace" windowpane with length 1000? (yes/no)
(4) Does the `numel` function also confirm the vector has length 1000? (yes/no)
(5) Include your plot here:

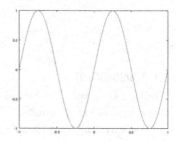

(6) Include your version of the file `exer2.m` here:

```
% Chapter 1, exercise 2
% A sample script file.
% M. Sussman
% $Id: introMatlab.tex,v 1.2 2021/05/22 15:36:04 mike Exp $

meshPoints=linspace(-1,1,1000);
meshPoints(95)      % results in -0.81181
numel(meshPoints) % results in 1000
plot(meshPoints,sin(2*pi*meshPoints))

% most students will probably take a screenshot
print -dpng exer2plot.png
```

When you test your script, does it give the same result as before? (yes/no)

Solution 1.3.

(1) Compute

```
colVec2 = (pi/4) * colVec1
```

```
colVec2 =
    1.5708
    7.0686
    6.2832
```

(2) Compute

```
colVec2 = cos( colVec2 )
colVec2 =
    6.1232e-17
    7.0711e-01
    1.0000e+00
```

Are these the values you expect? (yes/no), except some students may be surprised that cos(pi/2) is roundoff away from 0.

(3) Compute

```
colVec3 = colVec1 + colVec2
colVec3 =
    2.0000
    9.7071
    9.0000
```

(4) Compute norm(colVec3) = 13.388

(5) Compute

```
colvec4 = mat1 * colVec1
colvec4 =
    69
    95
    88
```

(6) Compute the following two expressions

```
>> mat1Transpose = mat1'
mat1Transpose =
    1    7    2
    3    9    4
    5    0    6
>> rowVec2 = colVec3'
rowVec2 =
    2.0000    9.7071    9.0000
```

(7) Compute the following four expressions

```
>> mat2 = mat1 * mat1'       % symmetric matrix
mat2 =
```

```
    35    34    44
    34   130    50
    44    50    56
>> rowVec3 = rowVec1 * mat1
rowVec3 =
  -47  -75  -59
>> dotProduct = colVec3' * colVec1
dotProduct =   163.36
>> euclideanNorm = sqrt(colVec2' * colVec2)
euclideanNorm =   1.2247
```

(8) Compute the following two expressions

```
>> determinant = det( mat1 )
determinant = -22.000
>> tr = trace( mat1 )
tr =   16
```

(9) Compute

```
>> min(rowVec1)
ans = -9
```

(10) Compute

```
>> max(mat1)
ans =
    7    9    6
```

(11) Compute the max norm of a vector.

```
>> max(abs(rowVec1))
ans =   9
```

(12) How would you find the single largest element of a matrix?
    `max(max(matrix))`

(13) Verify the matrix A is a magic square: All of the following sums are the same.

```
>> A=magic(201);
>> min(sum(A))
ans =   4060401
>> max(sum(A))
ans =   4060401
>> min(sum(A'))
ans =   4060401
>> max(sum(A'))
ans =   4060401
```

```
>> sum(diag(A))
ans =   4060401
>> sum(diag(fliplr(A)))
ans =   4060401
```

(14) Compute the following four expressions

```
>> integers = 0 : 10
integers =
    0    1    2    3    4    5    6    7    8    9   10
>> squareIntegers = integers .* integers
squareIntegers =
    0    1    4    9   16   25   36   49   64   81  100
>> cubeIntegers = squareIntegers .* integers
cubeIntegers =
    0    1    8   27   64  125  216  343  512  729 1000
>> tableOfPowers=[integers', squareIntegers', cubeIntegers']
tableOfPowers =
     0      0      0
     1      1      1
     2      4      8
     3      9     27
     4     16     64
     5     25    125
     6     36    216
     7     49    343
     8     64    512
     9     81    729
    10    100   1000
```

(15) Compute the following expressions

```
>> sqIntegers = integers .^ 2
sqIntegers =
    0    1    4    9   16   25   36   49   64   81  100

norm(sqIntegers-squareIntegers) = 0
```

(16) Compute the following two expressions

```
>> tableOfCubes = tableOfPowers(:,[1,3])
tableOfCubes =
     0      0
     1      1
     2      8
     3     27
```

4	64
5	125
6	216
7	343
8	512
9	729
10	1000

```
>> tableOfEvenCubes = tableOfPowers(1:2:end,[1:2:3])
tableOfEvenCubes =
```

0	0
2	8
4	64
6	216
8	512
10	1000

(17) What commands would be needed to generate the four 5×5 matrices in the upper left quarter AUL AUL=A(1:5,1:5)
the upper right quarter AUR AUR=A(1:5,6:10)
the lower left quarter ALL ALL=A(6:10,1:5)
and the lower right quarter ALR ALR=A(6:10,6:10)

(18) Is norm(A-B) zero? (yes/no)

(19) Compute

```
>> surprise = colVec1 + rowVec1
surprise =
    1  -2  -7
    8   5   0
    7   4  -1
```

Solution 1.4.

(1) Use cut-and-paste to put the code directly into the MATLAB command window to execute it. What is the final value for approxIntegral? 1.7184
Is it nearly $e^1 - e^0 = 1.7183$? (yes/no)

(2) Nothing required.

(3) What is the complete sequence of all values taken on by the variable x?
-h, 0, h, 2*h, ..., 40*h

(4) How many times is the statement executed? 2 (k==1 and k==N)

(5) How many times is the statement executed? (N-2)=38

Solution 1.5.

(1) Nothing required.

(2) One sentence on the relation between x and x.

The variable x is an array of sample values representative of the values $-1 \leq x \leq 1$.

(3) One sentence:

The line beginning y=y performs one term of the summation.

(4) What does the line do?

It creates an array y of the same size as x but filled with the value 0.

(5) Include the plot here:

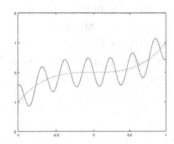

(6) What would happen if the two lines were omitted?

Only the green line would remain in the plot.

Solution 1.6.

(2) Include your code for **exer6.m** here:

```
% Chapter 1, exercise 6
% File named exer6.m
% M. Sussman

% compute terms of the Fourier Series for y=x^3 until converged
% plot the result using NPOINTS points from -1 to 1.
% Your name and the date

TOLERANCE=0.05; % the chosen tolerance value
NPOINTS=1000;
x=linspace(-1,1,NPOINTS);
y=zeros(size(x));
k=0;
term=TOLERANCE+1;
while max(abs(term)) > TOLERANCE
   k=k+1;
   term=2*(-1)^(k+1)*(pi^2/k-6/k^3)*sin(k*x);
   y=y+term;
end
```

```
disp( strcat('Number of iterations=',num2str(k)) )
plot(x,y,'b');   % 'b' is for blue line
hold on
plot(x,x.^3,'g'); % 'g' is for a green line
axis([-1,1,-2,2]);
hold off
```

(3) What is the purpose of the statement
So that the value of `term` is larger than TOLERANCE at the beginning.
(4) What is the purpose of the statement
So that the value of k increases as the loop progresses.
(5) How many iterations are required? 395
Does it generate a plot similar to the one from Exercise 6? (yes/no) (with smaller oscillations about the $y = x^3$ line.

Solution 1.7.

(1) Include your `exer7.m` here:

```
function k = exer7( tolerance )
  % k = exer7( tolerance )
  % compute terms of the Fourier Series for y=x^3 until converged
  % to the specified tolerance
  % k is the number of terms required
  % M. Sussman

  NPOINTS=1000;
  x=linspace(-1,1,NPOINTS);
  y=zeros(size(x));
  k=0;
  term=tolerance+1;
  while max(abs(term)) > tolerance
    k=k+1;
    term=2*(-1)^(k+1)*(pi^2/k-6/k^3)*sin(k*x);
    y=y+term;
  end
end
```

(2) What is the result of the command `exer7(0.05)`? Prints 395
(3) Use the command **help** `exer7` to display your help comments here:

```
k = exer7( tolerance )
compute terms of the Fourier Series for y=x^3 until converged
to the specified tolerance
k is the number of terms required
M. Sussman
```

(4) How many iterations are required for a tolerance of 0.05? 395

Does this value should agree with the value you saw in Exercise 6? (yes/no)

(5) How many iterations are required for tolerance of 0.1? 198

For 0.05? 395

For 0.025? 784

For 0.0125? 1580

Solution 1.8.

(1) Include your exer8.m here:

```
function [y,k] = exer8( tolerance )
  % [y,k] = exer8( tolerance )
  % compute terms of the Fourier Series for y=x^3 until converged
  % to the specified tolerance
  % y is a vector of length 1000 representing the number of
  % points in the interval [-1,1]
  % k is the number of terms required to reach the tolerance
  % M. Sussman

  NPOINTS=1000;
  x=linspace(-1,1,NPOINTS);
  y=zeros(size(x));
  k=0;
  term=tolerance+1;
  while max(abs(term)) > tolerance
     k=k+1;
     term=2*(-1)^(k+1)*(pi^2/k-6/k^3)*sin(k*x);
     y=y+term;
  end
end
```

(2) Command line would return *only* the number of iterations?

```
[~,n]=exer8(0.05);
```

Does the number of iterations agree with previous results? (yes/no)

(3) Command line that returns *only* the sum (a vector)?

```
z=exer8(0.05);
%     OR
[z,~]=exer8(0.05);
```

What is the norm of this vector, using **format long** 11.9962008192632

(4) Command line would return *both* the sum *and* the number of iterations for a tolerance of 0.02?

```
[z,n]=exer8(0.05);
```

How many iterations? 987
Norm of the sum, using `format long`? 11.9909579361475

Solution 1.9.

(1) Include your `exer9.m` here:

```
function [y,k] = exer9( tolerance, func )
  % [y,k] = exer9( tolerance, func )
  % compute terms of a series for y=x^3 in terms of func(k*x)
  % until converged to the specified tolerance
  % y is a vector of length 1000 representing the number of
  % points in the interval [-1,1]
  % k is the number of terms required to reach the tolerance
  % M. Sussman

  NPOINTS=1000;
  x=linspace(-1,1,NPOINTS);
  y=zeros(size(x));
  k=0;
  term=tolerance+1;
  while max(abs(term)) > tolerance
    k=k+1;
    term=2*(-1)^(k+1)*(pi^2/k-6/k^3)*func(k*x);
    y=y+term;
  end
end
```

(2) Is `norm(y8-y9)` small? (yes/no)
(3) What is the result of the command?

```
Error using sin
Not enough input arguments.
```

(4) Include the plot here:

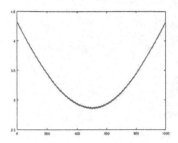

Chapter 2

Roots of equations

Solution 2.1. Plot the functions $y = \cos x$ and $y = x$ together in the same graph from $-\pi$ to π.

(1) ```
xplot=linspace(-pi,pi,200);
y1plot=cos(xplot);
y2plot=xplot;
plot(xplot,y1plot)
hold on
plot(xplot,y2plot)
hold off
print -dpng exercise1.png %NOT REQUIRED
```
(2) Include a copy of your plot here:

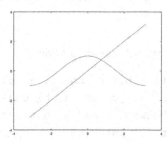

(3) What is the approximate value of $x$ at the intersection point? x=0.74.

**Solution 2.2.**

(1) Include your copy of cosmx.m here:

```
function y = cosmx ( x )
  % y = cosmx(x) computes the difference y=cos(x)-x

  % M. Sussman

  y = cos(x)-x;

end
```

15
```

(2) `cosmx(0.5)=0.37758`
(3) Include your plot here:

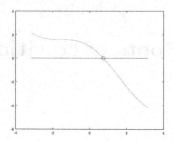

(4) Does the value of x at the point where the curve crosses the x-axis agree with
the previous exercise? (yes/no)
The point (0.74,0) is indicated with a circle on the plot. (not required)

Solution 2.3.

(1) The formula for the number of steps required by the bisection method in terms
of a and b and ϵ is $-\log_2(\epsilon/(|b - a|))$ rounded up to the nearest integer.
(2) Give an example of a continuous function that is equal to zero only once in the
interior of the interval [-2, 1], but for which bisection could not be used.
$f(x) = x^2$ or any non-negative function with a multiple root.

Solution 2.4.

(1) Include your code for `bisect_cosmx.m` here:

```
function [x,itCount] = bisect_cosmx( a, b)
  % [x,itCount] = bisect_cosmx( a, b) uses bisection to find a
  % root of cosmx between a and b to tolerance of 1.0e-10
  % a=left end point of interval
  % b=right end point of interval
  % cosmx(a) and cosmx(b) should be of opposite signs
  % x is the approximate root found
  % itCount is the number of iterations required.

  % your name and the date
  % $Id: roots.tex,v 1.2 2021/05/22 15:36:05 mike Exp $

  EPSILON = 1.0e-10;
  fa = cosmx(a);
  fb = cosmx(b);

  for itCount = 1:ceil(-log(EPSILON/(abs(b-a)))/log(2))
```

```
% from Exercise 3

  x = (b+a)/2;
  fx = cosmx(x);

  % The following statement prints the progress of the algorithm
  disp(strcat(  'a=' , num2str(a), ', fa=' , num2str(fa), ...
                ', x=' , num2str(x), ', fx=' , num2str(fx), ...
                ', b=' , num2str(b), ', fb=' , num2str(fb)));

  if ( fx == 0 )
     return;  % found the solution exactly!
  elseif ( abs ( b - x ) < EPSILON )
     return;  % satisfied the convergence criterion
  end

  if ( sign(fa) * sign(fx) <= 0 )
     b = x;
     fb = fx;
  else
     a = x;
     fa = fx;
  end

end

  error('bisect_cosmx failed with too many iterations!')
end
```

(2) Why is the name EPSILON all capitalized?
    To indicate that is a parameter and not intended to be changed.
(3) Is EPSILON related to eps? (yes/<u>no</u>)
    If so, how?
(4) In your own words, what does the sign(x) function do? What if x is 0?

    If x>0, sign(x)=1; if x<0, sign(x)=-1; if x==0, sign(x)=0;

(5) No response needed.
(6) What is the result if the MATLAB error function is called? The indicated error
    message is printed and execution is returned to the command prompt.
(7) What is the result of the command

    [z,iterations] = bisect_cosmx (0, 3)

    a=0, fa=1, x=1.5, fx=-1.4293, b=3, fb=-3.99

```
    ... many similar lines ...
a=0.7391,fa=4.634e-11,x=0.7391,fx=-9.9787e-11,b=.7391,
fb=-2.4591e-10, z =  0.73909
iterations =  35
```

Is z close to a root of the equation cos(z)=z? (yes/no)
(8) How many iterations? 35
    Is this value no larger than the value from your formula? (yes/no)
(9) What would itCount be? Undefined because the loop is never entered. *The point of this question is to cement in the student's mind that such a loop is never entered.*
(10) Include the result of the command help bisect_cosmx here:

```
[x,itCount] = bisect_cosmx( a, b) uses bisection to find a
root of cosmx between a and b to tolerance of 1.0e-10
a=left end point of interval
b=right end point of interval
cosmx(a) and cosmx(b) should be of opposite signs
x is the approximate root found
itCount is the number of iterations required.
```

(11) For what input numbers will this program produce an incorrect answer (*i.e.,* return a value that is not close to a root)? If the interval $[a, b]$ is not a change-of-sign interval.

   What code did you add to check that this cannot happen?

```
if sign(fa) * sign(fb) > 0
   error( ...
      'bisect_cosmx: sign(cosmx(a)) must be different from
      sign(cosmx(b))')
end
```

(12) Why is abs in the convergence test? Because $f(x) - f(b)$ might be negative, and that would satisfy the test even if its value were large.

Solution 2.5.

(1) Include your code for bisect.m here:

```
function [x,itCount] = bisect( func, a, b)
   % [x,itCount] = bisect( func, a, b) uses bisection to find a
   % func is a function handle for the function whose root is to
   % be found
   % root of func between a and b to tolerance of 1.0e-10
   % a=left end point of interval
   % b=right end point of interval
```

```
% func(a) and func(b) should be of opposite signs
% x is the approximate root found
% itCount is the number of iterations required.

% your name and the date

EPSILON = 1.0e-10;
fa = func(a);
fb = func(b);

if sign(fa) * sign(fb) > 0
  error( ...
  'bisect: sign(func(a)) must be different from sign(func(b))')
end

for itCount = 1:ceil(-log(EPSILON/(abs(b-a)))/log(2)) %from Ex. 3
    x = (b+a)/2;
    fx = func(x);

% The following statement prints the progress of the algorithm
    disp(strcat(  'a=' , num2str(a), ', fa=' , num2str(fa), ...
                 ', x=' , num2str(x), ', fx=' , num2str(fx), ...
                 ', b=' , num2str(b), ', fb=' , num2str(fb)));

    if ( fx == 0 )
       return;  % found the solution exactly!
    elseif ( abs ( b - x ) < EPSILON )
       return;  % satisfied the convergence criterion
    end

    if ( sign(fa) * sign(fx) <= 0 )
       b = x;
       fb = fx;
    else
       a = x;
       fa = fx;
    end

end
error('bisect failed with too many iterations!')
end
```

(2) Do you get the same result from the command

```
[z, iterations] = bisect ( @cosmx, 0, 3 )
```

as you got before from `bisect_cosmx`? (yes/no)

(3) Do you get the correct root of $f_0(x) = 1 - x$ within `EPSILON`? (yes/no)

Solution 2.6.

- Include your code for `f1.m` here:

```
function y = f1 ( x )
  % y = f1(x) computes x^2-9
  % M. Sussman
  y = x^2-9;
end
```

- Include your code for `f2.m` here:

```
function y = f2 ( x )
  % y = f2(x) computes x^5-x-1
  % M. Sussman
  y = x^5-x-1;
end
```

- Include your code for `f3.m` here:

```
function y = f3 ( x )
  % y = f3(x) computes x*exp(-x)
  % M. Sussman
  y = x*exp(-x);
end
```

- Include your code for `f4.m` here:

```
function y = f4 ( x )
  % y = f4(x) computes 2*cos(3*x)-exp(x)
  % M. Sussman
  y = 2*cos(3*x)-exp(x);
end
```

- Fill in the table

Name	Formula	Interval	approxRoot	No. Steps
f1	x^2-9	[0,5]	3.0000	36
f2	x^5-x-1	[1,2]	1.1673	34
f3	x*exp(-x)	[-1,2]	< EPSILON	35
f4	2*cos(3*x)-exp(x)	[0,6]	0.28203	36
f5	(x-1)^5	[0,3]	1.0000	35

Solution 2.7.

(1) Include your code for bisect0.m here:

```
function [x,itCount] = bisect0( func, a, b)
   % [x,itCount] = bisect0( func, a, b) uses bisection to find a
   % func is a function handle for the function whose root is sought
   % root of func between a and b to tolerance of 1.0e-10
   % a=left end point of interval
   % b=right end point of interval
   % func(a) and func(b) should be of opposite signs
   % x is the approximate root found  ∙
   % itCount is the number of iterations required.

   % M. Sussman

   EPSILON = 1.0e-10;
   fa = func(a);
   fb = func(b);

   if sign(fa) * sign(fb) > 0
      error('bisect0: sign(func(a)) must be opposite sign(func(b))')
   end

   for itCount = 1:10000
      x = (b+a)/2;
      fx = func(x);

      if ( fx == 0 )
         return; % found the solution exactly!
      elseif ( abs ( fx ) < EPSILON )
         return;  % satisfied the convergence criterion
      end

      if ( sign(fa) * sign(fx) <= 0 )
         b = x;
         fb = fx;
      else
         a = x;
         fa = fx;
      end
   end
   error('bisect0 failed with too many iterations!')
end
```

(2) Does bisect0 find the correct answer? (yes/no)
(3) For f0, the residual error is 5.8208e-11 and true error is 5.8208e-11.
    How many iterations did it take? 34
(4) For f5, the residual error is 2.9104e-11 and true error is 7.8125e-3.
    How many iterations did it take? 7
(5) To summarize your comparison, fill in the following table

Name	Formula	Interval	Residual error	True error	Number of steps
f0	x-1	[0,3]	5.8208e-11	5.8208e-11	34
f5	(x-1)^5	[0,3]	2.9104e-11	7.8125e-3	7

(6) Does the table show that the residual error is always smaller than EPSILON?
    (yes/no)
    What about the true error? (yes/no)

Solution 2.8.

(1) Include your code for secant.m here:

```
function [x,itCount] = secant( func, a, b)
  % [x,itCount] = secant( func, a, b) uses secant method to find a
  % func is a function handle for the function whose root is sought
  % root of func between a and b to tolerance of 1.0e-10
  % a=left end point of interval
  % b=right end point of interval
  % func(a) and func(b) should be of opposite signs
  % x is the approximate root found
  % itCount is the number of iterations required.

  % M. Sussman

  EPSILON = 1.0e-10;
  fa = func(a);
  fb = func(b);

  for itCount = 1:1000
    x = b - (b-a)*fb/(fb-fa);
    fx = func(x);

    if ( fx == 0 )
       return;  % found the solution exactly!
    elseif ( abs ( fx ) < EPSILON )
       return;  % satisfied the convergence criterion
    end
```

```
    a=b;
   fa=fb;
    b=x;
   fb=fx;
  end


   error('secant failed with too many iterations!')
  end
```

(2) Does it converge to the exact solution in 1 iteration? (yes/no)
    Why? For this linear function, the secant agrees with f0.
(3) Fill in the following table.

Name	Formula	Interval	Secant approxRoot	Secant Steps	Bisection approxRoot	Bisect Steps
f1	x^2-9	[0,5]	3.0000	7	3.0000	36
f2	x^5-x-1	[1,2]	1.1673	8	1.1673	34
f3	x*exp(-x)	[-1,2]	26.720	32	≈ 0	35
f4	2*cos(3*x) -exp(x)	[0,6]	0.28203	10	0.28203	36
f5	(x-1)^5	[0,3]	0.99065	31	1.0000	35

(4) Regarding the bisection roots as accurate, which functions are examples of
    convergence to a value that is not near a root in the table? f3
    Which are examples of inaccurate roots? f5

Solution 2.9.

(1) Include your code for regula.m here:

```
  function [x,itCount] = regula( func, a, b)
    % [x,itCount] = regula( func, a, b) uses regula falsi to find a
    % func is a function handle for the function whose root is sought
    % root of func between a and b to tolerance of 1.0e-10
    % a=left end point of interval
    % b=right end point of interval
    % func(a) and func(b) should be of opposite signs
    % x is the approximate root found
    % itCount is the number of iterations required.

    % your name and the date

    EPSILON = 1.0e-10;
    fa = func(a);
    fb = func(b);
```

```
    if sign(fa) * sign(fb) > 0
       error('regula: sign(func(a)) must be opposite sign(func(b))')
    end


    for itCount = 1:10000

       x = b-(b-a)*fb/(fb-fa);
       fx = func(x);

       if ( fx == 0 )
          return;  % found the solution exactly!
       elseif ( abs ( fx ) < EPSILON )
          return;  % satisfied the convergence criterion
       end

       if ( sign(fa) * sign(fx) >= 0 )
          a = b;
          fa = fb;
       end
       b=x;
       fb=fx;

    end

    error('regula failed with too many iterations!')
    end
```

(2) Test by finding the root of f0(x)=x-1 on the interval [-1,2]. Do you get the exact answer in a single iteration? (yes/no)

(3) Fill in the following table. Pay attention: the last line is for f3, not f5.

Name	Formula	Interval	Regula approxRoot	Regula Steps	Secant Steps	Bisection Steps
f1	x^2-9	[0,5]	3.0000	20	7	36
f2	x^5-x-1	[1,2]	1.1673	87	8	34
f3	x*exp(-x)	[-1,2]	≈0	56	32	35
f4	2*cos(3*x) -exp(x)	[0,6]	0.28203	278	10	36
f3	x*exp(-x)	[-1.5,1]	≈0	95	52	35

(4) Fill in the following additional line of the table.

Name	Formula	Interval	Regula approxRoot	Regula No. Steps
f5	(x-1)^5	[0,3]	0.93690	242,232

(5) Why would it be wrong to use the same convergence criterion in `regula.m` as was used in `bisect.m`? The size of the interval *does not necessarily* halve each iteration.

Solution 2.10.

(1) Include your code for `muller.m` here:

```
function [x2, itCount] = muller( func, a, b)
  % [x,itCount] = muller( func, a, b) uses Muller's method
  % func is a function handle for the function whose root is sought
  % root of func between a and b to tolerance of 1.0e-10
  % a=left end point of interval
  % b=right end point of interval
  % func(a) and func(b) should be of opposite signs
  % x is the approximate root found
  % itCount is the number of iterations required.

  % M. Sussman

  EPSILON = 1.0e-10;


  x0 = a;
  x2 = b;
  x1 = 0.51*x0 + 0.49*x2;
  y0=func(x0);
  y2=func(x2);
  y1=func(x1);


  for itCount = 1:100


    % Coefficients of quadratic passing through 3 points
    A = ( (y0 - y2) * (x1 - x2) - (y1 - y2) * (x0 - x2) ) / ...
            ( (x0 - x2) * ( x1- x2) * (x0 - x1) );
    B = ( (y1 - y2) * (x0 - x2)^2 - (y0 - y2) * (x1 - x2)^2 ) / ...
            ( (x0 - x2) * (x1 - x2) * (x0 - x1) );
    C = y2;


    % Find roots of quadratic polynomial if real and pick one
    % closer to x2
```

```
    if A ~= 0

        disc = B*B - 4.0*A*C;
        disc = max( disc, 0.0 );

        q1 = (B + sqrt(disc) );
        q2 = (B - sqrt(disc) );

        if abs(q1) < abs(q2)
            dx = -2.0*C/q2;
        else
            dx = -2.0*C/q1;
        end

    elseif B ~= 0
        dx = -C/B;
    else
        error(['muller: algorithm broke down at itCount=', ...
                num2str(itCount)])
    end

    x0 = x1;
    y0 = y1;

    x1 = x2;
    y1 = y2;

    x2 = x1 + dx;
    y2 = func(x2);

    if ( y2 == 0 )
        return;  % found the solution exactly!
    elseif ( abs ( y2 ) < EPSILON )
        return;  % satisfied the convergence criterion
    end
end
```

(2) Test on $f_0(x) = x - 1$ starting with the interval [0,2]. Do you observe convergence in a single iteration? (yes/no) Explain why it requires only a single iteration. The quadratic function through the three points x0, x1, x2 agrees with the linear function f0, so the root is found immediately.

(3) Test on f1 on the interval [0, 5]. Do you observe convergence in a single iteration? (yes/no)

Explain why it requires only a single iteration. Again, the fitted quadratic agrees with `f1` so the root is found immediately.

(4) Fill in the following table.

Name	Interval	Muller approxRoot	Muller Steps	Regula Steps	Secant Steps	Bisection Steps
f1	[0,5]	3.0000	1	13	7	36
f2	[1,2]	1.1673	7	53	8	34
f3	[-1,2]	46.503	25	36	32	35
f4	[0,6]	0.28203	8	171	10	36
f5	[0,3]	0.99242	13	242,232	31	35

Chapter 3

The Newton–Raphson method

Solution 3.1.

(1) Formulæ for derivatives of functions

f0	y=x-1	yprime=1
f1	y=x^2-9	yprime=2*x
f2	y=x^5-x-1	yprime=5*x^4-1
f3	y=x*exp(-x)	yprime=(1-x)*exp(-x)
f4	y=2*cos(3*x)-exp(x)	yprime=-6*sin(3*x)-exp(x)

(2) Include the code for f2.m here:

```
function [y,yprime]=f2(x)
  % [y,yprime]=f1(x) computes y=x-1 and its derivative, yprime=1

  % M. Sussman
  if numel(x)>1 % check that x is a scalar
    error('f1: x must be a scalar!')
  end

  y=x^5-x-1;
  yprime=5*x^4-1;
end
```

(3) Results of `help` commands.

Result of `help f0`
    [y,yprime]=f1(x) computes y=x-1 and its derivative, yprime=1
Result of `help f1`
    [y,yprime]=f1(x) computes y=x^2-9 and its derivative, yprime=2*x
Result of `help f2`
    [y,yprime]=f1(x) computes y=x-1 and its derivative, yprime=1

29

Result of `help f3`

```
[y,yprime]=f3(x) computes y=x*exp(-x) and
its derivative, yprime=(1-x)*exp(-x)
```

Result of `help f4`

```
[y,yprime]=f4(x) computes y=2*cos(3*x)-exp(x) and
its derivative, yprime=-6*sin(3*x)-exp(x)
```

(4) Function and derivative values

Function	function value at $x = -1$	derivative value at $x = -1$
f0	-2	1
f1	-8	-2
f2	-1	4
f3	-2.7183	5.4366
f4	-2.3479	0.47884

Solution 3.2.

(1) Writing `newton.m` Include your code for `newton.m` here:

```
function [x,numIts]=newton(func,x,maxIts)
  % [x,numIts]=newton(func,x,maxIts)
  % func is a function handle with signature [y,yprime]=func(x)
  % on input, x is the initial guess
  % on output, x is the final solution
  % EPSILON is convergence criterion = 5.0e-5
  % maxIts is largest number of iterates taken
  % maxIts is an optional argument
  % the default value of maxIts is 100
  % numIts is number of Iterations taken so far
  % Newton's method is used to find x so that func(x)=0

  % M. Sussman

  % check that x is a scalar
  if numel(x) > 1
    error('newton: x must be a scalar')
  end
  if nargin < 3
```

```
    maxIts=100;      % default value if maxIts is omitted
end


% convergence criterion
EPSILON = 5.0e-5;

for numIts=1:maxIts
   [value,derivative]=func(x);
   increment=-value/derivative;
   x = x + increment;
   errorEstimate = abs(increment);

   disp(strcat(num2str(numIts), ' x=', num2str(x), ...
      ' error estimate=', num2str(errorEstimate)));

   if errorEstimate<EPSILON
      return;
   end
end
% if get here, the Newton iteration has failed!
error('newton: maximum number of iterations exceeded.')
end
```

(2) Testing your code

   (a) Result of help newton command

```
      [x,numIts]=newton(func,x,maxIts)
      func is a function handle with signature [y,yprime]=func(x)
      on input, x is the initial guess
      on output, x is the final solution
      EPSILON is convergence criterion = 5.0e-5
      maxIts is largest number of iterates taken
      maxIts is an optional argument
      the default value of maxIts is 100
      numIts is number of Iterations taken so far
      Newton's method is used to find x so that func(x)=0
```

   (b) approxRoot=1, numIts=2

(3) Checking your work

   (a) Number of iterations=3
   (b) Estimated limit of error ratios $r_2^{(k)}$ =.167
   (c) For $x =$ your approximate solution, $|x - 3|$ =8.8818e-15
       Is it roughly the same size as EPSILON? Smaller!
   (d) Starting from x=-0.1, is your solution x=-3? Yes

(4) Using your code

Name	Formula	guess	approxRoot	No. Steps
f0	x-1	10	1	2
f1	x^2-9	0.1	3.0000	9
f2	x^5-x-1	10	1.1673	14
f3	x*exp(-x)	0.1	-2.2221e-16	4
f4	2*cos(3*x)-exp(x)	0.1	0.28203	5
f4	2*cos(3*x)-exp(x)	1.5	-3.6694	11

Solution 3.3.

(1) Include your code for f6.m here:

```
function [y,yprime]=f6(x)
  % [y,yprime]=f6(x) computes y=(x-4)^2 and its derivative, yprime

  % your name and the date
  if numel(x)>1 % check that x is a scalar
    error('f6: x must be a scalar!')
  end

  y=(x-4)^2;
  yprime=2*(x-4);
end
```

(2) It took <u>3</u> iterations for newton to find a root of f1=x^2-9.
(3) It took <u>17</u> iterations for newton to find a root of f6, starting from x=0.1. The true error=abs(x-4)=<u>2.9755e-05</u> and is larger or <u>smaller</u> than EPSILON.
(4) Include your code for f7.m here:

```
function [y,yprime]=f7(x)
  % [y,yprime]=f7(x) computes y=(x-4)^2 and its derivative, yprime

  % your name and the date
  if numel(x)>1 % check that x is a scalar
    error('f7: x must be a scalar!')
  end

  y=(x-4)^20;
  yprime=20*(x-4)^19;
end
```

It takes <u>163</u> iterations to find the root. The true error is <u>9.1204e-04</u> and is <u>larger</u> or smaller than EPSILON

(5) $r_1 = \underline{0.95001}$, and $r_2 = \underline{1.8802e+04}$.

Solution 3.4.

(1) Coding changes
    Include your new version of **newton.m** here:

```
function [x,numIts]=newton(func,x,maxIts)
  % [x,numIts]=newton(func,x,maxIts)
  % func is a function handle with signature [y,yprime]=func(x)
  % on input, x is the initial guess
  % on output, x is the final solution
  % EPSILON is convergence criterion = 5.0e-5
  % maxIts is largest number of iterates taken
  % maxIts is an optional argument
  % the default value of maxIts is 100
  % numIts is number of Iterations taken so far
  % Newton's method is used to find x so that func(x)=0

  % M. Sussman

  % check that x is a scalar or a column vector
  if numel(x) > 1
    error('newton: x must be a scalar')
  end
  if nargin < 3
    maxIts=100;      % default value if maxIts is omitted
  end

  % convergence criterion
  EPSILON = 5.0e-5;

  increment=1;  % this is an arbitrary value
  for numIts=1:maxIts
    [value,derivative]=func(x);
    oldIncrement=increment;
    increment=-value/derivative;
    x = x + increment;

    r1=abs(increment)/abs(oldIncrement);      % Equation 3.5
    r2=abs(increment)/abs(oldIncrement)^2;    % Equation 3.6
    errorEstimate = abs(increment);
```

```
%disp(strcat(num2str(numIts), ' x=', num2str(x), ...
%   ' error estimate=',num2str(errorEstimate)));

disp(strcat(num2str(numIts), ' r1=', num2str(r1), ' r2=', ...
   num2str(r2)));

if errorEstimate<EPSILON
   return;
end
end
% if get here, the Newton iteration has failed!
error('newton: maximum number of iterations exceeded.')
end
```

(2) Using your code

Function	numIts	r1	r2	true error	True err smaller than est?
f1=x^2-9	9	0.00019618	0.16667	0	yes
f6=(x-4)^2	17	0.5	8402.051	2.9755e-05	yes
f7=(x-4)^20	163	0.95	18801.329	9.1204e-04	no

Solution 3.5.

(1) Include your revised version of `newton.m` here:

```
function [x,numIts]=newton(func,x,maxIts)
% [x,numIts]=newton(func,x,maxIts)
% func is a function handle with signature [y,yprime]=func(x)
% on input, x is the initial guess
% on output, x is the final solution
% EPSILON is convergence criterion = 5.0e-5
% maxIts is largest number of iterates taken
% maxIts is an optional argument
% the default value of maxIts is 100
% numIts is number of Iterations taken so far
% Newton's method is used to find x so that func(x)=0

% M. Sussman

% check that x is a scalar or a column vector
if numel(x) > 1
   error('newton: x must be a scalar')
end
```

```
if nargin < 3
  maxIts=100;      % default value if maxIts is omitted
end

% convergence criterion
EPSILON = 5.0e-5;

increment=1;   % this is an arbitrary value
for numIts=1:maxIts
  [value,derivative]=func(x);
  oldIncrement=increment;
  increment=-value/derivative;
  x = x + increment;
  r1=abs(increment)/abs(oldIncrement);      % Equation 3.5
  r2=abs(increment)/abs(oldIncrement)^2;    % Equation 3.6
  errorEstimate = abs(increment);

% disp(strcat(num2str(numIts), ' r1=', num2str(r1), ' r2=', ...
%    num2str(r2)));

  if errorEstimate < EPSILON*(1-r1)
    return;
  end
end
% if get here, the Newton iteration has failed!
error('newton: maximum number of iterations exceeded.')
end
```

(2) Using your code
    Fill in the following table

Function	numIts	true err	True err smaller than est?
f1=x^2-9	9	8.8818e-15	yes
f6=(x-4)^2	18	1.4877e-05	yes
f7=(x-4)^20		4.6557e-05	yes

Solution 3.6.

(1) Include your cosmx.m here:

```
function [y,yprime]=cosmx(x)
  % [y,yprime]=cosmx(x) computes y=cos(x)-x and derivative=yprime

  % your name and the date
  if numel(x)>1 % check that x is a scalar
    error('cosmx: x must be a scalar!')
  end

  y=cos(x)-x;
  yprime=-sin(x)-1;
end
```

(2) Root of cosmx is 0.73909. Starting from x=0.5, number of iterations is 4.
(3) It did/didn't fail to find a root of cosmx starting from x=12
(4) With maxIts=5000, it took 113 iterations. It did/didn't find the same root.
    Warning: The value 113 varies largely with version and platform! Another
    version and platform resulted in 236 iterations.
(5) The final estimated error is 1.6429e-07 and the penultimate estimated error is
    0.0008628. The final error is/isn't smaller than the square of the penultimate
    error. but the ultimate error and penultimate error squared are comparable.

Solution 3.7.

(1) Include your code for f8=x^2+9 here:

```
function [y,yprime]=f8(x)
  % [y,yprime]=f8(x) computes y=x^2+9 and its derivative, yprime

  % your name and the date
  if numel(x)>1 % check that x is a scalar
    error('f8: x must be a scalar!')
  end

  y=x^2+9;
  yprime=2*x;
end
```

(2) Briefly describe the behavior of the iterations:
    Iterations are highly erratic, but the final few iterations are rapidly convergent.

Solution 3.8.

Initial guess	numIts	true error
1+1i	6	5.7809e-17
1-1i	6	5.7809e-17
10+5i	5	8.9309e-18
10+eps*i	58	4.4412e-16

Solution 3.9.

(1) Writing the code

- Include your code for f9.m here:

```
function [y,yprime]=f9(z)
  % [y,yprime]=f9(z) computes y=z^3+1 and its derivative, yprime

  % your name and the date
  if numel(z)>1 % check that x is a scalar
    error('f9: x must be a scalar!')
  end

  y=z^3+1;
  yprime=3*z^2;
end
```

- What is the purpose of the statement clear i
  Restore i as the imaginary unit if it has been used otherwise.
- What would happen if the two statements hold on and hold off were omitted?
  Only one point would be retained on the plot, as the plot command constantly refreshes the frame.
- If the variable root happened to take on the value root=0.50-i*0.86, what would the values of difference and whichRoot be? (Recall that (sqrt(3)/2)=.866025....)
  difference=0.006025   whichRoot=3
- If the function f9 were not properly programmed, it might be possible for the variable root to take on the value root=0.51-i*0.88. If this happened, what would the values of difference and whichRoot be?
  difference=0.017184   whichRoot=4
- Include your final code here:

```
% create a fractal figure from convergence of Newton iterations
% from initial conditions on a square in the complex plane
% M. Sussman
NPTS=100;
clear i
```

```
clear which
x=linspace(-2,2,NPTS);
y=linspace(-2,2,NPTS);
omega(1)= -1;
omega(2)= (1+sqrt(3)*i)/2;
omega(3)= (1-sqrt(3)*i)/2;

close %causes current plot to close, if there is one
for k=1:NPTS
  for j=1:NPTS
    z=x(k)+i*y(j);
    %plot(z,'.k') % plot as black points
    root=newton(@f9,z,500);
    [difference,whichRoot]=min(abs(root-omega));
    if difference>1.e-2
      whichRoot=4;
    end
    which(k,j)=whichRoot;
  end
end

imghandle=image(x,y,which');
colormap('flag') % red=1, white=2, blue=3, black=4
axis square
set(get(imghandle,'Parent'),'YDir','normal')
% plot the three roots as blacK asterisks
hold on
plot(real(omega),imag(omega),'k*')
hold off
```

(2) (a) Include your plot here:

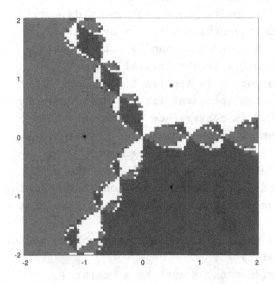

(b) (**Optional**) Include your second plot here:

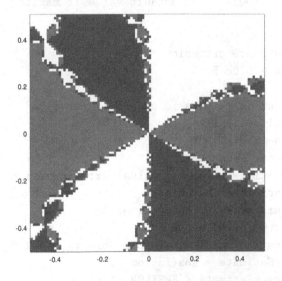

Solution 3.10.

(1) Include your code for `newtonfd.m` here:

```
function [x,numIts]=newtonfd(func,x,dx,maxIts)
  % [x,numIts]=newtonfd(func,x,dx,maxIts)
  % func is a function handle with signature [y,yprime]=func(x)
  % on input, x is the initial guess
  % on output, x is the final solution
  % dx is the increment for the derivative by finite differences
  % EPSILON is convergence criterion = 5.0e-5
  % maxIts is largest number of iterates taken
  % maxIts is an optional argument
  % the default value of maxIts is 100
  % numIts is number of Iterations taken so far
  % Newton's method is used to find x so that func(x)=0

  % check that x is a scalar or a column vector
  if numel(x) > 1
    error('newton: x must be a scalar')
  end
  if nargin < 4
    maxIts=100;      % default value if maxIts is omitted
  end

  % convergence criterion
  EPSILON = 5.0e-5;

  increment=1;  % this is an arbitrary value
  for numIts=1:maxIts
    value=func(x);
    valuenext=func(x+dx);
    derivative=(valuenext-value)/dx; % Equation 3.7
    oldIncrement=increment;
    increment=-value/derivative;
    x = x + increment;
    r1=abs(increment)/abs(oldIncrement);    % Equation 3.5
    errorEstimate = abs(increment);
    if errorEstimate < EPSILON*(1-r1)
      return;
    end
  end
  % if get here, the Newton iteration has failed!
  error('newtonfd: maximum number of iterations exceeded.')
end
```

(2) Testing your code

What is the converged value? <u>3</u>

How many iterations? <u>9</u>

Are these essentially the same as for `newton`? (<u>Yes</u>/No)

(3) Using your code

	Number of steps `f2`	Number of steps `f6`
using newton	11	16
dx = 0.00001	11	16
dx = 0.0001	11	16
dx = 0.001	11	24
dx = 0.01	11	113
dx = 0.1	14	1012
dx = 0.5	28	5011

Solution 3.11.

(1) Comparison of the two expressions

$$x \leftarrow x^{(k+1)}$$

$$a \leftarrow x^{(k-1)}$$

$$b \leftarrow x^{(k)}$$

$$\frac{f(b) - f(a)}{b - a} \approx f'(x^{(k)})$$

(2) Include your code for `newsecant.m` here:

```
function [x,numIts]=newsecant(func,x,maxIts)
  % [x,numIts]=newsecant(func,x,maxIts)
  % func is a function handle with signature [y,yprime]=func(x)
  % on input, x is the initial guess
  % on output, x is the final solution
  % EPSILON is convergence criterion = 5.0e-5
  % maxIts is largest number of iterates taken
  % maxIts is an optional argument
  % the default value of maxIts is 100
  % numIts is number of Iterations taken so far
  % Secant method is used to find x so that func(x)=0
  % additional initial starting value a=x-0.1;

  % M. Sussman

  % check that x is a scalar or a column vector
  if numel(x) > 1
     error('newsecant: x must be a scalar')
```

```
end
if nargin < 3
  maxIts=100;      % default value if maxIts is omitted
end

% convergence criterion
EPSILON = 5.0e-5;

increment=1;   % this is an arbitrary value
a=x-0.1;
fa=func(a);
b=x;
for numIts=1:maxIts
  fb=func(b);
  oldIncrement=increment;
  increment=-(b-a)/(fb-fa)*fb;
  a=b;
  fa=fb;
  b = b + increment;
  r1=abs(increment)/abs(oldIncrement);      % Equation 3.5
  errorEstimate = abs(increment);
  if errorEstimate < EPSILON*(1-r1)
     x=b;
     return;
  end
end
% if get here, the secant iteration has failed!
error('newsecant: maximum number of iterations exceeded.')
end
```

(3) Fill in the following table.

Function	start	Newton numIts	Newton solution	newsecant numIts	newsecant solution
f1=x^2-9	0.1	9	3.0000	13	3.0000
f3=x*exp(-x)	0.1	4	-2.2221e-16	2	0
f6=(x-4)^2	0.1	18	4.0000	25	4.0000
f7=(x-4)^20	0.1	221	4.0000	318	4.0000

Solution 3.12.

(1) Explain why (3.11) in the text is the result of applying Newton's method to the function $f(x) = x^2 - a$.

$$x^{(k+1)} = x^{(k)} - f(x^{(k)})/f'(x^{(k)})$$

$$= x^{(k)} - \frac{(x^{(k)})^2 - a}{2x^{(k)}}$$

$$= \frac{x^{(k)} + a}{2x^{(k)}}$$

(2) In the case $a > 0$, explain why convergence cannot deteriorate from quadratic. For $a > 0$, $\sqrt{a} > 0$.

(3) Include your MATLAB code for newton_sqrt.m here:

```
function [x,numIts]=newton_sqrt(a,maxits)
  % [x,numIts]=newton_sqrt(x,maxIts)
  % a is the number whos square root is needed
  % x is the square root of a
  % maxIts is largest number of iterates taken
  % maxIts is an optional argument
  % the default value of maxIts is 100
  % numIts is number of Iterations taken so far
  % Equation 3.9 is the solution method

  % M. Sussman
  % check that a is a scalar
  if numel(a) > 1
     error('newton_sqrt: x must be a scalar')
  end
  if nargin < 2
     maxIts=100;      % default value if maxIts is omitted
  end

  x=.5*(1+a);
  for numIts=1:maxIts
    oldx=x;
    x=.5*(x+a/x);
    if abs(x-oldx)<= 2*eps*x
       return
    end
  end
  % if get here, the secant iteration has failed!
  error('newton_sqrt: maximum number of iterations exceeded.')
end
```

(4) Fill in the following table

Value	square root	true error	no. iterations
a=9	3	0	6
a=1000	31.623	0	10
a=12345678	3513.6	0	16
a=0.000003	0.0017321	0	14

Remark 3.1. It is unusual to see true errors of exactly zero in ordinary computations. The computer chips used in most PC and Mac computers have square roots computed in hardware. The precise algorithms used vary with compiler and application options as well as particular chip model. In this case, the exact agreement between hardware and Newton iteration is most likely due to the rapid convergence rate of Newton.

Chapter 4

Multidimensional Newton's method

Solution 4.1.

(1) Include your code for `vnewton.m` here:

```
function [x,numIts]=vnewton(func,x,maxIts)
% [x,numIts]=vnewton(func,x,maxIts)
% func is a function handle with signature [y,yprime]=func(x)
% if x is a column vector, y must also be a column vector of
%    the same size and yprime must be a square matrix
% on input, x is the initial guess
% on output, x is the final solution
% EPSILON is convergence criterion = 5.0e-5
% maxIts is largest number of iterates taken
% maxIts is an optional argument
% the default value of maxIts is 100
% numIts is number of Iterations taken so far
% Newton's method is used to find x so that func(x)=0

% M. Sussman

% check that x is a scalar or a column vector
[rows,cols]=size(x);
if cols>1
   error('vnewton: x must be a scalar or column vector')
end
if nargin < 3
   maxIts=100;     % default value if maxIts is omitted
end

% convergence criterion
EPSILON = 5.0e-5;
```

```
    increment=1;   % this is an arbitrary value
    for numIts=1:maxIts
      [value,derivative]=func(x);
      oldIncrement=increment;
      increment=-derivative\value;
      x = x + increment;
      r1=norm(increment)/norm(oldIncrement);      % Equation 3.5
      errorEstimate = norm(increment);

      if errorEstimate < EPSILON*(1-r1)
        return;
      end
    end
    % if get here, the Newton iteration has failed!
    error('vnewton: maximum number of iterations exceeded.')
  end
```

(2) Your solution is <u>-5.7809e-17 + 3.0000e+00i</u>, and <u>numIts=6</u>. Do they agree with the result from newton? <u>yes</u>/no
(3) Fill in the following table,

Initial guess	numIts	abs(error)	numIts(newton)
1+i	6	5.7809e-17	6
1-i	6	5.7809e-17	6
10+5i	7	1.3130e-12	7
10+eps*i	61	3.4532e-12	61

Solution 4.2.

(1) Include your code for f8v.m here:

```
function [f,J]=f8v(x)
  % [f,J]=f8v(x) computes y=x^2+9 and its Jacobian matrix, J

  % M. Sussman

  % check that x is a 2-dimensional column vector
  [rows,cols]=size(x);
  if cols>1 | rows ~= 2
    error('f8v: x must be a 2-dimensional column vector!')
  end

  % f is to be a column vector
```

```
f= [ x(1)^2-x(2)^2+9
        2*x(1)*x(2)  ];
df1dx1= 2*x(1);
df1dx2=-2*x(2);
df2dx1= 2*x(2);
df2dx2= 2*x(1);
J= [ df1dx1   df1dx2
        df2dx1   df2dx2 ];
end
```

(2) Does your solution using vnewton.m using f8v.m agree with the results in Exercise 4.1? yes/no

Are the numbers of iterations the same? yes/no

(3) Fill in the table below.

Initial guess	numIts	norm(error)
[1;1]	6	5.7809e-17
[1;-1]	6	5.7809e-17
[10;5]	7	1.3130e-12
[10;eps]	61	3.4532e-12

Solution 4.3.

(1) Include your plot here: (or include it as a separate file)

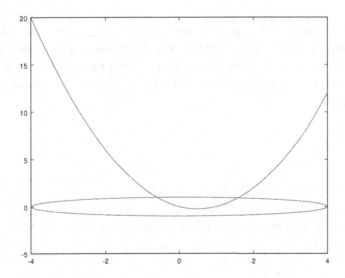

(2) Include your code for f3v.m here:

```
function [f,J]=f3v(x)
```

```
% [f,J]=f3v(x)
% f and x are both 2-dimensional column vectors
% J is a 2 X 2 matrix
% f(1) = x(1)^2 - x(1) - x(2)  ,  f(2) = x(1)^2/16 + x(2)^2 -1

% M. Sussman

% check that x is a 2-dimensional column vector
[rows,cols]=size(x);
if cols>1 | rows ~= 2
  error('f8v: x must be a 2-dimensional column vector!')
end

% f is to be a column vector
f= [ ( x(1)^2 - x(1) - x(2) )
      ( x(1)^2/16 + x(2)^2 -1 ) ];
df1dx1= 2*x(1) - 1;
df1dx2=-1;
df2dx1=x(1)/8;
df2dx2=2*x(2);
J= [ df1dx1   df1dx2
      df2dx1   df2dx2 ];
end
```

(3) Starting from the initial column vector guess [2;1], the intersection point that
vnewton finds is [1.58101; 0.91857] and the number of iterations is 4.

(4) What initial guess did you use to find the other intersection point? [-2;1], what
is the other intersection point [-0.61274;0.98820], and how many iterations did
it take? 5

Solution 4.4.

(1)–(4) Include your code for vnewton1.m here:

    (5) Does your test using f3v.m yield the same results as Exercise 4.3? (yes/no)
Does the matrix iterations have 2 rows and (numIts+1) columns?
(yes/no)

Solution 4.5.

(2) Include a copy of your plot here:

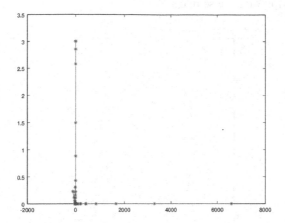

(3) Include a copy of your zoomed plot here:

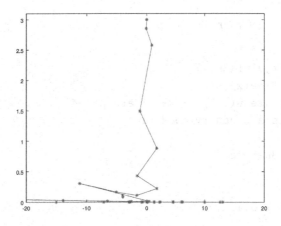

(4) Explain why if $x_2 = 0$ initially, $x_2 = 0$ for all subsequent iterations.
From f8v,

```
f= [ x(1)^2-x(2)^2+9
        2*x(1)*x(2)  ];
df1dx1= 2*x(1);
df1dx2=-2*x(2);
df2dx1= 2*x(2);
df2dx2= 2*x(1);
```

So, if $x_2 = 0$, then $f_2 = 0$ and $df_1/dx_2 = 0$, $df_2/dx_1 = 0$ so that $J = 2x_1 I$, where I is the identity matrix. Since $f_2 = 0$, the next iterate will have $x_2 = 0$ and also all subsequent iterates, by induction.

(6) Include a copy of your semilog plot here

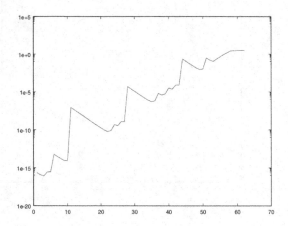

Solution 4.6.

(1) Include a copy of your code here:

```
function [f,J]=flow(x)
  % [f,J]=flow(x)
  % x(1-7) are u(1-7), x(8-11) are p(1-4)
  % u(0) is a boundary condition

  % M. M. Sussman

  u0=0.1e-3;
  K=[1,3,5,7,9,11];
  u=x(1:7);
  p=x(8:11);
  f(1 ,1) = u0-u(1)-u(4)-u(5);
  f(2 ,1) = u(1)-u(2)-u(6);
  f(3 ,1) = u(2)-u(3)+u(5);
  f(4 ,1) = u(3)+u(4)+u(6)-u(7);
  f(5 ,1) = p(2)-p(1) -K(1)*u(1)*abs(u(1));   %x(9), x(8)
  f(6 ,1) = p(3)-p(2) -K(2)*u(2)*abs(u(2));   %x(10), x(9)
  f(7 ,1) = p(4)-p(3) -K(3)*u(3)*abs(u(3));   %x(11), x(10)
  f(8 ,1) = p(4)-p(1) -K(4)*u(4)*abs(u(4));   %x(11), x(8)
  f(9 ,1) = p(3)-p(1) -K(5)*u(5)*abs(u(5));   %x(10), x(8)
  f(10,1) = p(4)-p(2) -K(6)*u(6)*abs(u(6));   %x(11), x(9)
  f(11,1) = p(1)+p(2)+p(3)+p(4);              % x(8), 9, 10, 11
```

```
J(1 ,:) = [                    -1, 0, 0,-1,-1, 0, 0, 0, 0, 0, 0];
J(2 ,:) = [                     1,-1, 0, 0, 0,-1, 0, 0, 0, 0, 0];
J(3 ,:) = [                     0, 1,-1, 0, 1, 0, 0, 0, 0, 0, 0];
J(4 ,:) = [                     0, 0, 1, 1, 0, 1,-1, 0, 0, 0, 0];
J(5 ,:) = [-K(1)*2*abs(u(1)), 0, 0, 0, 0, 0, 0,-1, 1, 0, 0];
J(6 ,:) = [ 0,-K(2)*2*abs(u(2)), 0, 0, 0, 0, 0, 0,-1, 1, 0];
J(7 ,:) = [ 0, 0,-K(3)*2*abs(u(3)), 0, 0, 0, 0, 0, 0,-1, 1];
J(8 ,:) = [ 0, 0, 0,-K(4)*2*abs(u(4)), 0, 0, 0,-1, 0, 0, 1];
J(9 ,:) = [ 0, 0, 0, 0,-K(5)*2*abs(u(5)), 0, 0,-1, 0, 1, 0];
J(10,:) = [ 0, 0, 0, 0, 0,-K(6)*2*abs(u(6)), 0, 0,-1, 0, 1];
J(11,:) = [ 0, 0, 0, 0, 0,                    0, 0, 1, 1, 1, 1];
end
```

(2) Does your Jacobian matrix check out? (yes/no)

Solution 4.7.

(1) The final three values of `norm(increment)` are

```
iteration=1, norm(increment)=4.122918e+00
iteration=2, norm(increment)=3.726845e+00
iteration=3, norm(increment)=9.387965e-02
iteration=4, norm(increment)=2.999838e-03
iteration=5, norm(increment)=1.644873e-06
```

(2) Does your solution satisfy $u_7 = u_0$? (yes/no)
(3) $u_4 = $ 9.9555e-01
(4) $u_1 = $ 0.36577, $u_2 = $ 0.36414, $u_3 = $ 0.36577

Solution 4.8.

(3) Is f at x=[0.15;2.0;1.0;3] zero? (yes/no)
(4) Is the determinant of J at x=[0.15;2.0;1.0;3] non-zero?   (yes/no)
   (det(J)=2.4402e+11)
(5) Is J symmetric? (yes/no) (norm(J-J')=0)
   What are the four eigenvalues of J? 2.8680e+01, 3.8729e+01, 1.4816e+04, 1.4828e+04

Solution 4.9.

Initial guess	Number of iterations
[0.15; 2.0; 1.0; 3]	1
[0.15; 2.0; 0.9; 3]	3
[0.15; 2.0; 0.0; 3]	6
[0.15; 2.0;-0.1; 3]	6
[0.15; 2.0;-0.3; 3]	7
[0.15; 2.0;-0.5; 3]	failed
[0.15; 2.0; 1.0; 4]	5
[0.15; 2.0; 1.0; 5]	6
[0.15; 2.0; 1.0; 6]	9
[0.15; 2.0; 1.0; 7]	failed
[0.15; 1.99; 1.0; 3]	3
[0.15; 1.97; 1.0; 3]	4
[0.15; 1.95; 1.0; 3]	5
[0.15; 1.93; 1.0; 3]	7
[0.15; 1.91; 1.0; 3]	failed
[0.17; 2.0; 1.0; 3]	5
[0.19; 2.0; 1.0; 3]	7
[0.20; 2.0; 1.0; 3]	8
[0.21; 2.0; 1.0; 3]	failed

Solution 4.10.

(1) Include a copy of snewton0.m here:

```
function [x,numIts]=snewton0(func,x,maxIts)
  % [x,numIts]=snewton0(func,x,maxIts)
  % func is a function handle with signature [y,yprime]=func(x)
  % if x is a column vector, y must also be a column vector of
  %    the same size and yprime must be a square matrix
  % on input, x is the initial guess
  % on output, x is the final solution
  % EPSILON is convergence criterion = 5.0e-5
  % maxIts is largest number of iterates taken
  % maxIts is an optional argument
  % the default value of maxIts is 100
  % numIts is number of Iterations taken so far
  % Softened Newton's method with fixed parameter=0.5
  %    is used to find x so that func(x)=0

  % M. Sussman
```

```
% check that x is a scalar or a column vector
[rows,cols]=size(x);
if cols>1
  error('vnewton: x must be a scalar or column vector')
end
if nargin < 3
  maxIts=100;      % default value if maxIts is omitted
end

% convergence criterion
EPSILON = 5.0e-5;

% fixed softening parameter
ALPHA=0.5;

increment=1;  % this is an arbitrary value
for numIts=1:maxIts
  [value,derivative]=func(x);
  oldIncrement=increment;
  increment=-derivative\value;
  x = x + ALPHA*increment;
  r1=norm(increment)/norm(oldIncrement);      % Equation 3.5
  errorEstimate = norm(increment);

  if errorEstimate < EPSILON*(1-r1)
    return;
  end
end
% if get here, the softened Newton iteration has failed!
error('snewton0: maximum number of iterations exceeded.')
end
```

(2) How large can the first component of the initial guess get before the iteration diverges? 0.21

Solution 4.11.

(1) Include a copy of snewton1.m here:

```
function [x,numIts]=snewton1(func,x,maxIts)
  % [x,numIts]=snewton0(func,x,maxIts)
  % func is a function handle with signature [y,yprime]=func(x)
  % if x is a column vector, y must also be a column vector of
```

```
%     the same size and yprime must be a square matrix
% on input, x is the initial guess
% on output, x is the final solution
% EPSILON is convergence criterion = 5.0e-5
% maxIts is largest number of iterates taken
% maxIts is an optional argument
% the default value of maxIts is 100
% numIts is number of Iterations taken so far
% Softened Newton's method using Equation 4.7
%     to find x so that func(x)=0

% M. Sussman

% check that x is a scalar or a column vector
[rows,cols]=size(x);
if cols>1
  error('vnewton: x must be a scalar or column vector')
end
if nargin < 3
  maxIts=100;      % default value if maxIts is omitted
end

% softening parameter
BETA=10;

% convergence criterion
EPSILON = 5.0e-5;

increment=1;  % this is an arbitrary value
for numIts=1:maxIts
  [value,derivative]=func(x);
  oldIncrement=increment;
  increment=-derivative\value;
  alpha=1/(1+BETA*norm(increment)); % Equation 4.7
  x = x + alpha*increment;
  r1=norm(increment)/norm(oldIncrement);     % Equation 3.5
  errorEstimate = norm(increment);

  if errorEstimate < EPSILON*(1-r1)
    return;
  end
end
```

```
% if get here, the softened Newton iteration has failed!
error('snewton1: maximum number of iterations exceeded.')
end
```

(2) How many iterations are required to converge, starting from x(1)=0.20? 7
   How many iterations were required by snewton0? 16
(3) How large can the first component of the initial guess get before the iteration
   diverges? x=0.24

Solution 4.12.

(1) Include a copy of easy_objective.m here:

```
function [f,J,F]=easy_objective(x,x0)
  % [f,J,F]=easy_objective(x-x0)
  % more comments

  % your name and the date

  if norm(size(x)-size(x0)) ~= 0
     error('easy_objective: x and x0 must be compatible.')
  end

  F=sum((x-x0).^2);

  % f(k)=derivative of F with respect to x(k)
  f=zeros(4,1);
  f=2*(x-x0);

  % J(k,ell)=derivative of f(k) with respect to x(ell)
  J=diag([2,2,2,2]);
end
```

(2) For x0=[0;2;1;2] and x=[0;0;0;0], f=[0;-4;-2;-4]
   J=[2 0 0 0
   0 2 0 0
   0 0 2 0
   0 0 0 2]
   F=9
(3) For x0=[0;2;1;2] and x=[1;-1;1;-1], f=[-2;2;-2;2]
   J=[2 0 0 0
   0 2 0 0
   0 0 2 0
   0 0 0 2]
   F=4

(4) Include a copy of `homotopy.m` here:

```
function [f,J,F]=homotopy(x,p,x0)
% [f,J,F]=homotopy(x,p,x0)
% computes the homotopy or Davidenko objective function
% for 0<=p<=1

% M. Sussman

[f1,J1,F1]=objective(x);
[f2,J2,F2]=easy_objective(x,x0);
f=p*f1+(1-p)*f2;
J=p*J1+(1-p)*J2;
F=p*F1+(1-p)*F2;
end
```

(5) Does `dvdnko.m` successfully reach the value p=1? (yes/no)

(6) Does it get the same solution values for x as Exercise 4.9? (yes/no)
(7) Explain what `@(xx) homotopy(xx,p,x0)` means.
   It defines a new (unnamed) function with a single parameter that calls the homotopy function using the current values of p and x0.
   Why is this used instead of simply using `@homotopy`?
   The vnewton function requres a function with a single variable.
(8) Does `dvdnko` successfully reach p=1? (yes/no)
   How far can you increase the first component and still have it successfully reach p=1? 0.50 (fails at 0.55, cannot get past p=0.001)
(9) For `STEPS=750`, how far can you increase the first component and still have it successfully reach p=1? 0.25
(10) For `STEPS=1000`, can you start from x0=`[0;2;1;3]` and reach the solution? (yes/no)
   How about x0=`[-0.5;2;1;3]`? (yes/no)

Solution 4.13.

(1) Symbolic hand calculations for f and J are:

$$f_1 = (d_1 + \varepsilon)x_1^n - \frac{x_2^n}{4} - \frac{x_3^n}{9} - \frac{1}{6}$$

$$f_2 = -x_1^n + (d_2 + \varepsilon)x_2^n - \frac{x_3^n}{9} - \frac{2}{9}$$

$$f_3 = -\frac{x_1^n}{4} - x_2^n + (d_3 + \varepsilon)x_3^n - \frac{1}{4}$$

$$J = n \begin{bmatrix} (d_1 + \varepsilon)x_2^{n-1} & -x_2^{n-1}/4 & -x_3^{n-1}/9 \\ -x_1^{n-1} & (d_2 + \varepsilon)x_2^{n-1} & -x_3^{n-1}/9 \\ -x_1^{n-1}/4 & -x_2^{n-1} + (d_3 + \varepsilon)x_3^{n-1} \end{bmatrix}.$$

(2) A copy of your f13.m follows:

```
function [f,J]=f13(x)
  % [f,J]=f13(x)
  % large function to test quasi-Newton methods

  % your name and the date

  [N,M]=size(x);
  if M ~= 1
     error(['f13: x must be a column vector'])
  end

  n=2;  % exponent in function definition (4.9)
  epsilon=1.e-5; % epsilon in (4.9)
  f=zeros(N,1);
  J=zeros(N,N);

  j=(1:N)';
  jn=j.^n;
  dd=sum(1./jn);
  xn=x.^n;
  for k=1:N
     f(k)=(dd-1/k^2+epsilon)*xn(k) - sum(xn(k+1:N)./jn(k+1:N)) ...
          -sum(xn(k-1:-1:1)./jn(1:k-1))-k/(N*(k+1));
  end

  if nargout>1 % if need Jacobian too, compute it
     J=zeros(N,N);
     for k=1:N
        J(k,k)=(dd-1/k^2+epsilon)*n*x(k)^(n-1);
        J(k,k-1:-1:1)=-1./jn(1:k-1)*n.*x(k-1:-1:1).^(n-1);
        J(k,k+1:N)=-1./jn(k+1:N)*n.*x(k+1:N).^(n-1);
     end
  end
end
```

(3) Do your hand calculations agree with f13.m results for x=[1;1;1]? (yes/no)

(4) Include the matrix Ffd here:

```
  0.72224  -1.00000  -0.66667
 -2.00000   4.44448  -0.66667
 -0.50000  -4.00000   7.50006
```

Is norm(J - Jfd) zero or roundoff? (yes/no)

Solution 4.14.

(1) Include a copy of qnewton.m here:

```
function [x,numIts]=qnewton(func,x,maxIts)
  % [x,numIts]=qnewton(func,x,maxIts)
  % func is a function handle with signature [y,yprime]=func(x)
  % if x is a column vector, y must also be a column vector of
  %    the same size and yprime must be a square matrix
  % on input, x is the initial guess
  % on output, x is the final solution
  % EPSILON is convergence criterion = 5.0e-5
  % maxIts is largest number of iterates taken
  % maxIts is an optional argument
  % the default value of maxIts is 100
  % numIts is number of Iterations taken so far
  % quasi-Newton method is used to find x so that func(x)=0
  % with previous-step Jacobian used when convergence is fast enough

  % M. Sussman

  % check that x is a scalar or a column vector
  [rows,cols]=size(x);
  if cols>1
    error('qnewton: x must be a scalar or column vector')
  end
  if nargin < 3
    maxIts=100;      % default value if maxIts is omitted
  end

  % convergence criterion
  EPSILON = 5.0e-5;

  skipNext = false;
  increment=1;  % this is an arbitrary value
  for numIts=1:maxIts
    [value,derivative]=func(x);
    if ~skipNext
      tim=clock;
      Jinv=inv(derivative);
      inversionTime=etime(clock,tim);
    else
      inversionTime=0;
```

```
      end
      oldIncrement=increment;
      increment = -Jinv*value;
      x = x + increment;
      r1=norm(increment)/norm(oldIncrement);     % Equation 3.5

      % Use r1 to determine if the NEXT iteration should be skipped
      skipNext = r1 < 0.2;
      fprintf('it=%d, r1=%e, inversion time=%e.\n',numIts,r1, ...
            inversionTime)
      errorEstimate = norm(increment);

      if errorEstimate < EPSILON*(1-r1)
        return;
      end
    end
    % if get here, the Newton iteration has failed!
    error('qnewton: maximum number of iterations exceeded.')
  end
```

(2) Include a copy of **qnewtonNoskip.m** here:

```
function [x,numIts]=qnewtonNoskip(func,x,maxIts)
  % [x,numIts]=qnewtonNoskip(func,x,maxIts)
  % func is a function handle with signature [y,yprime]=func(x)
  % if x is a column vector, y must also be a column vector of
  %    the same size and yprime must be a square matrix
  % on input, x is the initial guess
  % on output, x is the final solution
  % EPSILON is convergence criterion = 5.0e-5
  % maxIts is largest number of iterates taken
  % maxIts is an optional argument
  % the default value of maxIts is 100
  % numIts is number of Iterations taken so far
  % Newton method is used to find x so that func(x)=0
  % uses explicit inverse for comparison with qnewton

  % M. Sussman

  % check that x is a scalar or a column vector
  [rows,cols]=size(x);
  if cols>1
    error('qnewtonNoskip: x must be a scalar or column vector')
```

```
   end
   if nargin < 3
     maxIts=100;        % default value if maxIts is omitted
   end

   % convergence criterion
   EPSILON = 5.0e-5;

   skipNext = false;
   increment=1;   % this is an arbitrary value
   for numIts=1:maxIts
     [value,derivative]=func(x);
     if ~skipNext
       tim=clock;
       Jinv=inv(derivative);
       inversionTime=etime(clock,tim);
     else
       inversionTime=0;
     end
     oldIncrement=increment;
     increment = -Jinv*value;
     x = x + increment;
     r1=norm(increment)/norm(oldIncrement);      % Equation 3.5

     skipNext = false;
     fprintf('it=%d, r1=%e, inversion time=%e.\n',numIts,r1, ...
             inversionTime)
     errorEstimate = norm(increment);

     if errorEstimate < EPSILON*(1-r1)
       return;
     end
   end
   % if get here, the Newton iteration has failed!
   error('qnewtonNoskip: maximum number of iterations exceeded.')
end
```

(3) qnewton: how long? <u>134.494</u> (depends)

How many iterations? <u>9</u>

How many iterations were skipped? <u>5</u>

What is the total time for inversion as a percentage of the total time taken?
<u>77%</u>

(4) qnewtonNoskip: how long? (204.416) (depends)
How many iterations? <u>7</u>
(norm(v-v0)/norm(v)=<u>1.8727e-09</u>

(5) Which is faster, qnewton or qnewtonNoskip? <u>qnewton</u>

Interpolation on evenly-spaced points

Solution 5.1.

(1) Include your code coef_vander.m below:

```
function c = coef_vander ( xdata, ydata )
  % c = coef_vander ( xdata, ydata )
% computes polynomial interpolation coefficients using Vandermonde
  % xdata= abscissas for interpolation points
  % ydata= ordinates for interpolation points
  % c= coefficients of interpolating polynomial

  % M. Sussman

  N=numel(xdata);
  if N ~= numel(ydata)
     error('coef_vander: xdata and ydata must have same length')
  end

  for j = 1:N
    for k = 1:N
       A(j,k) = xdata(j)^(N-k) ;
    end
  end
  c = A \ reshape(ydata,N,1);
end
```

(2) The coefficients are [1;0;0]. Are they correct? (yes/no) ($y = x^2 + 0x + 0$)
(3) The coefficients are [1;-2;3;-4;5;-6;7]
(4) What yval values correspond to the xdata values?
   1636, 247, 28, 7, 4, 31, 412.
   Are they correct? (yes/no)
(5) What polyfit command did you use? polyval(c,xdata)

Solution 5.2.

(1) cTrue = [1,-3,-5,15,4,-12]
(2) The value at x = 0 is 12 (constant term)
(3) cVander = [1;-3;-5;15;4;-12]
(4) What are the coefficients using only five roots? All zeros! The unique polynomial of degree 5 with 5 roots is trivial.
(5) Include your exer2.m here:

```
% construct many test points (for plotting)
xval=linspace(-2,3,4001);
% construct the true test point values, for reference
yvalTrue=polyval(cTrue,xval);

% use Vandermonde polynomial interpolation coefficients
% to evaluate the interpolant at the test points
yval=polyval(cVander, xval);

% plot reference values in thick green
plot(xval,yvalTrue,'g','linewidth',4);
hold on
% plot interpolation data points as black plus signs
plot(xdata,ydata,'k+');
% plot interpolant in thin black
plot(xval,yval,'k');
hold off

% estimate the relative approximation error of the interpolant
approximationError=max(abs(yvalTrue-yval))/max(abs(yvalTrue))
```

The difference is a roundoff number.

Solution 5.3.

(1) Include a copy of your exer3.m here:

```
% M. Sussman

% construct N=5 data points
N=5;
xdata=linspace(-pi,pi,N);
ydata=sinh(xdata);

% construct many test points (for plotting)
xval=linspace(-pi,pi,4001);
```

```
% construct the true test point values, for reference
yvalTrue=sinh(xval);

% use Vandermonde polynomial interpolation coefficients
% to evaluate the interpolant at the test points
% compute
cVander=coef_vander(xdata,ydata);
% evaluate
yval=polyval(cVander, xval);

% plot reference values in thick green
plot(xval,yvalTrue,'g','linewidth',4);
hold on
% plot interpolation data points as black plus signs
plot(xdata,ydata,'k+');
% plot interpolant in thin black
plot(xval,yval,'k');
hold off

% estimate the relative approximation error of the interpolant
approximationError=max(abs(yvalTrue-yval))/max(abs(yvalTrue))
```

(2) Does the hyperbolic sine and its interpolant agree at the interpolation points?
(<u>yes</u>/no)
(3) Fill in the following table:

N	Approximation error
5	0.038807
11	6.6116e-06
21	5.5373e-15

Solution 5.4.

(1) Include your runge.m here:

```
function y=runge(x)
  % y=runge(x)
  % computes the Runge example function

  % M. Sussman

  y=1./(1+x.^2);
end
```

(2) Include your `exer4.m` here:

```
% M. Sussman

% construct N=5 data points
N=5;
xdata=linspace(-pi,pi,N);
ydata=runge(xdata);

% construct many test points (for plotting)
xval=linspace(-pi,pi,4001);
% construct the true test point values, for reference
yvalTrue=runge(xval);

% use Vandermonde polynomial interpolation coefficients
% to evaluate the interpolant at the test points
% compute
cVander=coef_vander(xdata,ydata);
% evaluate
yval=polyval(cVander, xval);

% plot reference values in thick green
plot(xval,yvalTrue,'g','linewidth',4);
hold on
% plot interpolation data points as black plus signs
plot(xdata,ydata,'k+');
% plot interpolant in thin black
plot(xval,yval,'k');
hold off

% estimate the relative approximation error of the interpolant
approximationError=max(abs(yvalTrue-yval))/max(abs(yvalTrue))
```

(3) Does the Runge example function and its interpolant agree at the interpolation points? (yes/no)

(4) Fill in the following table

N	Approximation error
5	0.31327
11	0.58457
21	3.8607

(5) Are you surprised to see that the errors do not decrease? (yes/no) This question is intended to force the student to think about those results.

Solution 5.5.

(1) Fill in the following table.

N		Coefficients		
5	c_5 1	c_3 -0.35387	c_1 0.02653	
11	c_{11} 1	c_9 -0.89399	c_7 0.50096	c_5 -0.13984
21	c_{21} 1	c_{19} -0.99765	c_{17} 0.96078	c_{15} -0.80173
limit	+1	-1	+1	-1

(2) Fill in the following table.

N	max(abs(c(2:2:end)))
5	0
11	0
21	1.6148e-14

Solution 5.6.

(1) Explain in a sentence or two.

Clearly, the function f_j is zero at $x = x_j$ and 1 at $x = x_k$. Since ℓ_k contains all factors except the kth, it is zero for $j \neq k$, and one at $j = k$.

(2) Include a copy of your **lagrangep.m**

```
function pval = lagrangep( k, xdata, xval)
  % pval = lagrangep( k, xdata, xval)
  % computes the k-th Lagrange polynomial base on xdata at
  % the points xval
  % pval is the result of the calculation

  % M. Sussman

  N=numel(xdata);
  pval = 1;
  for j = 1:N
    if j ~= k
       pval = pval .* (xval - xdata(j)) / (xdata(k)-xdata(j));
    end
  end
end
```

(3) **lagrangep(1, xdata, xval) = [1 0 0]**
(4) Does **lagrangep** give the correct values for **lagrangep(1, xdata, xdata)**? (yes/no)

For **lagrangep(2, xdata, xdata)**? (yes/no)
For **lagrangep(3, xdata, xdata)**? (yes/no)

Solution 5.7.

(1) Include your `eval_lagr.m` here:

```
function yval = eval_lagr ( xdata, ydata, xval)
  % yval = eval_lagr ( xdata, ydata, xval)
  % evaluates the Lagrange interpolating polynomial based on xdata
  % and ydata at the points xval

  % M. Sussman

  N = numel(ydata);
  yval = 0;
  for k=1:N
     yval = yval + ydata(k)*lagrangep(k,xdata,xval);
  end
end
```

(2) Do you get `ydata` back? (yes/no)
(3) What are the values of the interpolant at `xval=xdata`?
    [1636 247 28 7 4 31 412]
(4) Repeat Exercise ?? using Lagrange interpolation. Include your `exer7.m` here:

```
% M. Sussman

% construct many test points (for plotting)
xval=linspace(-2,3,4001);
% construct the true test point values, for reference
yvalTrue=polyval(cTrue,xval);

% use Vandermonde polynomial interpolation coefficients
% to evaluate the interpolant at the test points
yval=eval_lagr(xdata,ydata, xval);

% plot reference values in thick green
plot(xval,yvalTrue,'g','linewidth',4);
hold on
% plot interpolation data points as black plus signs
plot(xdata,ydata,'k+');
% plot interpolant in thin black
plot(xval,yval,'k');
hold off

% estimate the relative approximation error of the interpolant
```

```
approximationError=max(abs(yvalTrue-yval))/max(abs(yvalTrue))
```

Include your plot here:

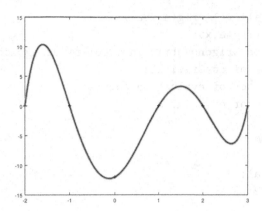

What is the error? 1.5399e-15

Solution 5.8.

(1) Include your `coef_trig.m` here:

```
function a=coef_trig(func,N)
  % a=coef_trig(func,N)
  % computes the trigonometric interpolation coefficients of func
  % func = a function handle
  % N= number of trig powers (number of terms = 2*N+1)
  % a=vector of coefficients

  % M. Sussman

  xdata = 2*pi*(-N:N)/(2*N+1);

  for k=1:2*N+1
    a(k)=sum( exp(-i*(k-N-1)*xdata).*func(xdata))/(2*N+1);
  end

end
```

(2) Which of the a_k are nontrivial? `a(12)`
What are the values? `a(12)=1`
(3) Which of the a_k are nontrivial? `a(7)`, `a(15)`
What are the values? `a(7)=0.5i`, `a(15)=-0.5i`

Solution 5.9.

(1) Include your `eval_trig.m` here:

```
function fval=eval_trig(a,xval)
  % a=eval_trig(a,xval)
  % evaluates trigonometric interpolant using coefficients a
  % a=vector of coefficients
  % xval=vector of evaluation points
  % fval=vector of results

  % M. Sussman

  NN=numel(a);
  N=(NN-1)/2;

  fval=0;
  for k=1:NN
    fval=fval+ a(k).*exp(i*(k-N-1).*xval);
  end
end
```

(2) Include your `exer9.m` here:

```
% M. Sussman

% degree of trig polynomial (number of terms = 2*N+1)
N=10;
% use a function handle for convenience
func=@(x) sin(4*x);

% construct many test points (for plotting)
xval=linspace(-2,3,4001);
% construct the true test point values, for reference
yvalTrue=func(xval);

% use trigonometric polynomial interpolation
% to evaluate the interpolant at the test points
c=coef_trig(func,N);
yval=eval_trig(c, xval);

% plot reference values in thick green
plot(xval,yvalTrue,'g','linewidth',4);
hold on
```

```
% plot interpolant in thin black
plot(xval,yval,'k');
hold off
```

```
% estimate the relative approximation error of the interpolant
approximationError=max(abs(yvalTrue-yval))/max(abs(yvalTrue))
```

Include your plot here:

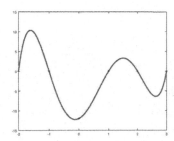

What is the approximation error? 2.1094e-15

Solution 5.10.

(1) Fill in the following table.

N	y = x.*(pi^2-x.^2) approximation error
5	0.011932
10	0.0032219
15	0.0014736
20	8.4129e-04

(2) Fill in the following table.

N	Runge approximation error
2	0.093216
4	0.021412
5	0.0059001

(3) What are the five points?

```
xdata = [ -2.51327  -1.25664   0.00000   1.25664   2.51327 ]
ydata = [  0.13668   0.38773   1.00000   0.38773   0.13668 ]
```

Do runge and eval_trig agree at those points?
(yes/no)
(norm(eval_trig(c,xdata)-ydata) is roundoff)

Solution 5.11.

(1) Include your `sawshape5.m` here:

```
function y=sawshape5(x)
  % y=sawshape5(x)
  % computes the sawtooth function
  % x=input
  % y=result

  % M. Sussman

  kless=find(x<0);
  kgreater=find(x>=0);
  y(kless)=x(kless)+pi;
  y(kgreater)=x(kgreater)-pi;
end
```

(2) What is the result of: `find(x==c)` 3
    `find(x==h)` 8, 12
    `find(x>=g)` 7:13
(3) Fill in the following table:

N	sawshape5 approximation error
5	1.9962
10	1.9927
100	1.9262

Include your plot for N=5 here:

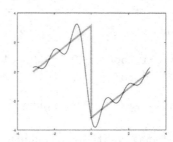

Include your plot for N=100 here:

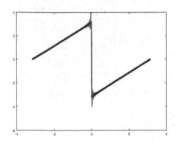

(4) Are the plus signs in the proper places? (yes/no)

Solution 5.12.

(1) Fill in the table.

ℓ_i	n_1	n_2	n_3
1	1	0	0
2	0	1	0
3	0	0	1

(2) Fill in the table.

q_i	n_1	n_2	n_3	n_4	n_5	n_6
1	1	0	0	0	0	0
2	0	1	0	0	0	0
3	0	0	1	0	0	0
4	0	0	0	1	0	0
5	0	0	0	0	1	0
6	0	0	0	0	0	1

(3) $x = 2\xi + \eta + 2$
(4) $y = 2\eta + 2$
(5) $x_0 = 2,\ y_0 = 2,$

$$J = \begin{bmatrix} 2 & 1 \\ 0 & 2 \end{bmatrix}$$

(6)

$$J^{-1} = \begin{bmatrix} 0.5 & -0.25 \\ 0 & 0.5 \end{bmatrix}$$

(7) What are the values a_k for $k = 1, \ldots, 6$? Fill in the table.

a_1	1.491825
a_2	2.225541
a_3	3.320117
a_4	1.822119
a_5	2.857651
a_6	2.117000

(8) $p(3.5, 2.5) = 2.412937$

Sample symbolic m-file follows:

```
% M. Sussman

clear
syms xi eta

half=sym(1)/sym(2);

ell1 = 1 - xi - eta;
ell2 = xi;
ell3 = eta;

q1 = 2*(1-xi-eta)*(half-xi-eta);
q2 = 2*xi*(xi-half);
q3 = 2*eta*(eta-half);
q4 = 4*xi*(1-xi-eta);
q5 = 4*xi*eta;
q6 = 4*eta*(1-xi-eta);

disp('(1) confirmed unless otherwise indicated.')
e(1)=subs(subs(ell1,xi,0),eta,0) - 1;
e(2)=subs(subs(ell2,xi,0),eta,0) - 0;
e(3)=subs(subs(ell3,xi,0),eta,0) - 0;
if sum(e.^2) > 0
    disp('(1) confirmation failed at n1')
end

e(1)=subs(subs(ell1,xi,1),eta,0) - 0;
e(2)=subs(subs(ell2,xi,1),eta,0) - 1;
e(3)=subs(subs(ell3,xi,1),eta,0) - 0;
if sum(e.^2) > 0
    disp('(1) confirmation failed at n2')
```

```
end

e(1)=subs(subs(ell1,xi,0),eta,1) - 0;
e(2)=subs(subs(ell2,xi,0),eta,1) - 0;
e(3)=subs(subs(ell3,xi,0),eta,1) - 1;
if sum(e.^2) > 0
    disp('(1) confirmation failed at n3')
end

disp('(2) confirmed unless otherwise indicated.')

e(1)=subs(subs(q1,xi,0),eta,0) - 1;
e(2)=subs(subs(q2,xi,0),eta,0) - 0;
e(3)=subs(subs(q3,xi,0),eta,0) - 0;
e(4)=subs(subs(q4,xi,0),eta,0) - 0;
e(5)=subs(subs(q5,xi,0),eta,0) - 0;
e(6)=subs(subs(q6,xi,0),eta,0) - 0;
if sum(e.^2) > 0
    disp('(2) confirmation failed at n1')
end

e(1)=subs(subs(q1,xi,1),eta,0) - 0;
e(2)=subs(subs(q2,xi,1),eta,0) - 1;
e(3)=subs(subs(q3,xi,1),eta,0) - 0;
e(4)=subs(subs(q4,xi,1),eta,0) - 0;
e(5)=subs(subs(q5,xi,1),eta,0) - 0;
e(6)=subs(subs(q6,xi,1),eta,0) - 0;
if sum(e.^2) > 0
    disp('(2) confirmation failed at n2')
end

e(1)=subs(subs(q1,xi,0),eta,1) - 0;
e(2)=subs(subs(q2,xi,0),eta,1) - 0;
e(3)=subs(subs(q3,xi,0),eta,1) - 1;
e(4)=subs(subs(q4,xi,0),eta,1) - 0;
e(5)=subs(subs(q5,xi,0),eta,1) - 0;
e(6)=subs(subs(q6,xi,0),eta,1) - 0;
if sum(e.^2) > 0
    disp('(2) confirmation failed at n3')
end

e(1)=subs(subs(q1,xi,half),eta,0) - 0;
```

```
e(2)=subs(subs(q2,xi,half),eta,0) - 0;
e(3)=subs(subs(q3,xi,half),eta,0) - 0;
e(4)=subs(subs(q4,xi,half),eta,0) - 1;
e(5)=subs(subs(q5,xi,half),eta,0) - 0;
e(6)=subs(subs(q6,xi,half),eta,0) - 0;
if sum(e.^2) > 0
    disp('(2) confirmation failed at n4')
end

e(1)=subs(subs(q1,xi,half),eta,half) - 0;
e(2)=subs(subs(q2,xi,half),eta,half) - 0;
e(3)=subs(subs(q3,xi,half),eta,half) - 0;
e(4)=subs(subs(q4,xi,half),eta,half) - 0;
e(5)=subs(subs(q5,xi,half),eta,half) - 1;
e(6)=subs(subs(q6,xi,half),eta,half) - 0;
if sum(e.^2) > 0
    disp('(2) confirmation failed at n5')
end

e(1)=subs(subs(q1,xi,0),eta,half) - 0;
e(2)=subs(subs(q2,xi,0),eta,half) - 0;
e(3)=subs(subs(q3,xi,0),eta,half) - 0;
e(4)=subs(subs(q4,xi,0),eta,half) - 0;
e(5)=subs(subs(q5,xi,0),eta,half) - 0;
e(6)=subs(subs(q6,xi,0),eta,half) - 1;
if sum(e.^2) > 0
    disp('(2) confirmation failed at n6')
end

disp('(3) find x(xi,eta) = eta + 2*xi +2 ')
x=simplify(2*ell1+4*ell2+3*ell3)

disp('(4) find y(xi,eta) = 2*eta +2')
y=simplify(2*ell1+2*ell2+4*ell3)

syms x0 y0 J
disp('(5) x0=2;y0=2;J=[2 1;0 2]')
x0=2;
y0=2;
J=[2 1;0 2];
xy=simplify( [x0;y0]+J*[xi;eta] )
if sum(simplify([x;y]-xy).^2) ~= 0
```

```
   disp('confirmation of vector x,y as function of xi,eta failed.')
end

disp('(6) compute xi1, eta1 as function of x,y')
Jinv=sym(inv(J/4));
Jinv=Jinv/4;
if sum((simplify(Jinv*[x-x0;y-y0] - [xi;eta])).^2) ~= 0
   disp('confirmation of vector xi,eta as function of x,y failed.')
end
% write xi,eta as functions of x1,y1
syms x1 y1
xieta= Jinv*[x1-x0;y1-y0]

% compute q's as functions of x1,y1
Q1=subs(subs(q1,xi,xieta(1)),eta,xieta(2));
Q2=subs(subs(q2,xi,xieta(1)),eta,xieta(2));
Q3=subs(subs(q3,xi,xieta(1)),eta,xieta(2));
Q4=subs(subs(q4,xi,xieta(1)),eta,xieta(2));
Q5=subs(subs(q5,xi,xieta(1)),eta,xieta(2));
Q6=subs(subs(q6,xi,xieta(1)),eta,xieta(2));

% turn Q into cue, a MATLAB function of x,y
cue1=matlabFunction(Q1);
cue2=matlabFunction(Q2);
cue=matlabFunction(Q3); % function of y1 only
cue3=@(x1,y1) cue(y1);   % make function of 2 variables
cue4=matlabFunction(Q4);
cue5=matlabFunction(Q5);
cue6=matlabFunction(Q6);

% double check cues
e(1)=cue1(2,2)   -1;
e(2)=cue1(4,2)   -0;
e(3)=cue1(3,4)   -0;
e(4)=cue1(3,2)   -0;
e(5)=cue1(3.5,3)-0;
e(6)=cue1(2.5,3)-0;
if sum(e.^2) > 0
    disp('cue confirmation failed at n1')
end

e(1)=cue2(2,2)   -0;
```

```
e(2)=cue2(4,2)  -1;
e(3)=cue2(3,4)  -0;
e(4)=cue2(3,2)  -0;
e(5)=cue2(3.5,3)-0;
e(6)=cue2(2.5,3)-0;
if sum(e.^2) > 0
    disp('cue confirmation failed at n2')
end

e(1)=cue3(2,2)  -0;
e(2)=cue3(4,2)  -0;
e(3)=cue3(3,4)  -1;
e(4)=cue3(3,2)  -0;
e(5)=cue3(3.5,3)-0;
e(6)=cue3(2.5,3)-0;
if sum(e.^2) > 0
    disp('cue confirmation failed at n3')
end

e(1)=cue4(2,2)  -0;
e(2)=cue4(4,2)  -0;
e(3)=cue4(3,4)  -0;
e(4)=cue4(3,2)  -1;
e(5)=cue4(3.5,3)-0;
e(6)=cue4(2.5,3)-0;
if sum(e.^2) > 0
    disp('cue confirmation failed at n4')
end

e(1)=cue5(2,2)  -0;
e(2)=cue5(4,2)  -0;
e(3)=cue5(3,4)  -0;
e(4)=cue5(3,2)  -0;
e(5)=cue5(3.5,3)-1;
e(6)=cue5(2.5,3)-0;
if sum(e.^2) > 0
    disp('cue confirmation failed at n5')
end

e(1)=cue6(2,2)  -0;
e(2)=cue6(4,2)  -0;
e(3)=cue6(3,4)  -0;
```

```
e(4)=cue6(3,2)   -0;
e(5)=cue6(3.5,3)-0;
e(6)=cue6(2.5,3)-1;
if sum(e.^2) > 0
    disp('cue confirmation failed at n6')
end

a(1)=exp(0.4);
a(2)=exp(0.8);
a(3)=exp(1.2);
a(4)=exp(0.6);
a(5)=exp(1.05);
a(6)=exp(0.75);
fprintf('(7) values of a are \n %f %f %f %f %f %f\n',a)

p=a(1)*cue1(3.5,2.5) + a(2)*cue2(3.5,2.5) + ...
  a(3)*cue3(3.5,2.5) + a(4)*cue4(3.5,2.5) + ...
  a(5)*cue5(3.5,2.5) + a(6)*cue6(3.5,2.5);
fprintf('(8) p(3.5,2.5)=%f\n',p)
```

Chapter 6

Polynomial and piecewise linear interpolation

Solution 6.1.

(1) Include your test_poly_interpolate.m here:

```
function max_error=test_poly_interpolate(func,xdata )
  % max_error=test_poly_interpolate(func,xdata )
  % utility function used for testing polynomial interpolation
  % func is the function to be interpolated
  % xdata are abscissae at which interpolation takes place
  % max_error is the maximum difference between the function
  %    and its interpolant

  % M. Sussman

  % Choose the number of the test points and generate them
  % Use 4001 because it is odd, capturing the interval midpoint
  NTEST=4001;
  % construct NTEST points evenly spaced so that
  % they cover the interpolation interval in a standard way, i.e.,
  % xval(1)=xdata(1) and xval(NTEST)=xdata(end)
  xval= linspace(xdata(1),xdata(end),NTEST);

  % we need the results of func at the points xdata to do the
  % interpolation WARNING: this is a vector statement
  % In a real problem, ydata would be "given" somehow, and
  % a function would not be available
  ydata=func(xdata);

  % use Lagrange interpolation from Chapter 5
  % WARNING: these use componentwise (vector) statements.
  % Generate yval as interpolated values corresponding to xval
```

81

```
yval=eval_lagr(xdata,ydata,xval);

% comparing yval with the exact results of func at xval
% In a real problem, the exact results would not be available.
yexact= func(xval);

% plot the exact and interpolated results on the same plot
% this gives assurance that everything is reasonable
plot(xval,yval,xval,yexact)

% compute the error in a standard way.
max_error=max(abs(yexact-yval))/max(abs(yexact));
end
```

(2) Does your code agree with the results? (yes/no)

(3) Fill in the following table.

Runge function, evenly spaced points $[-5, 5]$

ndata	Max Error
5	0.43836
11	1.9156
21	59.822

Solution 6.2.

(1) xdata=

```
-5.00000   -3.20000   -1.80000   -0.80000
-0.20000    0.00000    0.20000    0.80000
 1.80000    3.20000    5.00000
```

(2) Fill in the table

Runge function, points concentrated near 0

ndata	Max Error
5	1.9186
11	147.34
21	3.9559e+05

Solution 6.3.

(1) xdata=

```
-5.00000   -4.47214   -3.87298   -3.16228
```

-2.23607	0.00000	2.23607	3.16228
3.87298	4.47214	5.00000	

(2) Include your plot here:

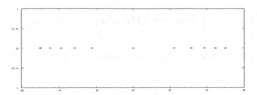

(3) Fill in the following table;

Runge function, points concentrated near endpoints

ndata	Max Error
5	0.45998
11	0.19938
21	0.061611

Solution 6.4.

(2) `rfactorial(5)=120`, `factorial(5)=120`, Are they the same? (yes/no)
(3) Explain how the computation continues to get a result of 6. The active copy of `rfactorial` has `n=1`, so it starts up another new copy of `rfactorial` for `n=0`. This copy returns 1, so now the version of `rfactorial` for `n=1` returns the value $1*1=1$, passing this value back to the version of `rfactorial` for `n=2`. This version, in turn computes $2*1=2$ and passes it back to the waiting version of `rfactorial` for `n=3`. This version returns $3*2=6$ back to the original caller.
(4) What would have happened if the code did not test if `n<0`? It would go on reducing the value of `n` into large negative values until the stack was full and the program crashed.

Solution 6.5.

(1) Include your `fibonacci.m` here:

```
function f=fibonacci(n)
  % f=fibonacci(n) computes n-th Fibonacci numbers recursively


  % M. Sussman
  % $Id: interp2.tex,v 1.2 2021/05/22 15:36:04 mike Exp $


  if n<=0
    error('fibonacci: cannot compute Fibonacci numbers for n<=0.');
  elseif n<=2
    f=1;
```

```
    else
      f=fibonacci(n-1)+fibonacci(n-2);
    end
  end
```

(2) Confirm that your function correctly computes the first few Fibonacci numbers: (yes/no)

(3) For n=13, fibonacci gives 233 and Binet's method gives 233.0

Solution 6.6.

(1) Include your cheby_trig.m here:

```
function tval=cheby_trig(xval,degree)
  % tval=cheby_trig(xval,degree),
  % trig definition of Chebyshev polynomial
  % degree=degree of desired polynomial, defaults to 7
  % xval=points at which polynomial is to be evaluated

  % M. Sussman

  if nargin==1
    degree=7;
  end

  tval=cos(degree*acos(xval));
end
```

(2) Include your cheby_recurs.m here:

```
function tval=cheby_recurs(xval,degree)
  % tval=cheby_recurs(xval,degree),
  % recursive definition of Chebyshev polynomial
  % degree=degree of desired polynomial, defaults to 7
  % xval=points at which polynomial is to be evaluated

  % M. Sussman

  if nargin==1
    degree=7;
  end

  if degree<0
    error('cheby_recurs: negative degree is not allowed.')
  elseif degree==0
```

```
      tval=ones(size(xval));
    elseif degree==1
      tval=xval;
    else
      tval=2*xval.*cheby_recurs(xval,degree-1) - ...
                    cheby_recurs(xval,degree-2);
    end
  end
```

(3) What is the value?

`max(abs(cheby_trig(xval,4)-cheby_recurs(xval,4)))` = 1.8190e-12

(4) Include your plot here:

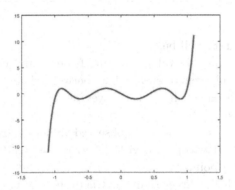

(5) What are the largest and second-largest roots of $T_7(x) = 0$ on the interval [-1.1, 1.1]? 0.97493, 0.78183

Solution 6.7.

(1) Include your `cheby_points.m` here:

```
function xdata = cheby_points( a, b, ndata)
  % xdata = cheby_points( a, b, ndata)
  % finds ndata Chebyshev points on the interval [a,b]

  % M. Sussman

  k = (1:ndata);
  theta=(2*k-1)*pi/(2*ndata);
  xdata=.5*(a+b+(a-b)*cos(theta));
end
```

(2) Do the largest and second largest Chebyshev points agree with your roots of T_7 computed above? (yes/no)

(3) What are the five Chebyshev points for **ndata=5** and **[a,b]=[-5,5]**?

       -4.7553e+00   -2.9389e+00   -3.0616e-16
        2.9389e+00   4.7553e+00

(4) Nothing required
(5) Fill in the table:

Runge function, Chebyshev points

ndata	Max Error
5	0.40202
11	0.10915
21	0.015334
41	2.8945e-04

Solution 6.8.

(2) Explain what **xdata** will be:
   a vector of 21 ascending values starting from -5 and ending with +5 with 19
   (pseudo-) randomly-generated numbers between them.
(3) Will you get the same values twice? (yes/<u>no</u>)
(4) Nothing required.
(5) What are the largest and smallest observed values of **err**? *random* 1.2445e+07
   0.047588 How do they compare with the error you found in the previous exercise
   for 21 Chebyshev points?
   The smallest value is larger than the Chebyshev point minimum.

Solution 6.9.

(1) Include your **scalar_bracket.m** here:

```
function left_index=scalar_bracket(xdata,xval)
  % left_index=scalar_bracket(xdata,xval)
  % left_index is the index so that xval is inside
  %      [xdata(left_index),xdata(left_index+1)]

  % M. Sussman

  ndata = numel(xdata);

  % first check Case 1
  if  xval < xdata(1)
    left_index = 1;
    return
  % then check Case 2
  elseif xdata(ndata) <= xval
```

```
    left_index = ndata-1;
    return
% finally check Case 3
else
  for k = 1:ndata-1
    if xdata(k) <= xval & xval < xdata(k+1)
      left_index = k;
      return
    end
  end
  error('Scalar_bracket: this cannot happen!')
  end
end
```

(2) Fill in the table with your values: *these choices are arbitrary but satisfy the requirements*

xdata	xval	left_indices (scalar)
1	0	1
2	6	4
3	3	3
4	1.5	1
5	4.5	4
—	1.01	1
—	4.99	4

Are all these values correct? (yes/no)

Solution 6.10.

(2) indices= [2 4 5 6]
(3) left_indices= [0 4 0 4 4 4]
(4) Explain the meaning of the statement beginning "if **any**"
   any is a MATLAB function that returns **true** if $\exists$ k so that left_indices(k) is zero.

   Explain why it indicates that "not all indices set."
   Since left_indices starts out as all zeros and is set to the value of a subscript (necessarily > 0) when given a value, if any zeros remain, then some index has been missed.
(5) Why do the two functions give the same results?
   In scalar_bracket, xval<xdata(1) and xdata(1)<=xval<xdata(2) both give left_index=1 but are checked separately. In bracket, they are checked once. Similarly for the right end point.

(6) Include your commented version of `bracket.m` here:
    Additional comments should explain the variables `xdata` and `xval`. Comments
    explaining the cases might be included, too.
(7) Summarize your results:

xdata	xval	left_indices (scalar)	left_indices (vector)
1	0	1	1
2	6	4	4
3	3	3	3
4	1.5	1	1
5	4.5	4	4
—	1.01	1	1
—	4.99	4	4

Solution 6.11.

(1) Include your `eval_plin.m` here:

```
function yval = eval_plin ( xdata, ydata, xval)
  % yval = eval_plin ( xdata, ydaa, xval)
  % evaluates a piecewise linear function passing through the
  % points (xdata(k), ydata(k)) at the values xval
  % yval is returned as a vector of the same shape as xval

  % M. Sussman

  left=bracket(xdata,xval);
  % y=(y_b*(x-a) + y_a*(b-x))/(b-a)
  % is linear and =y_b at a and y_a at b

  yval=(ydata(left+1).*(xval-xdata(left)  ) + ...
        ydata(left  ).*(xdata(left+1)-xval)) ./  ...
        (xdata(left+1) - xdata(left));
end
```

(2) Include your plot here:

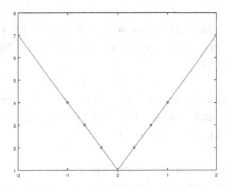

Solution 6.12.

(1) Include your `test_plin_interpolate.m` here:

```
function max_error=test_plin_interpolate(func,xdata )
  % max_error=test_plin_interpolate(func,xdata )
  % utility function used for testing piecewise linear interpolation
  % func is the function to be interpolated
  % xdata are abscissae at which interpolation takes place
  % max_error is the maximum difference between the function
  %   and its interpolant

  % M. Sussman

  % Choose the number of the test points and generate them
  % Use 4001 because it is odd, capturing the interval midpoint
  NTEST=4001;
  % construct NTEST points evenly spaced so that
  % they cover the interpolation interval in a standard way, i.e.,
  % xval(1)=xdata(1) and xval(NTEST)=xdata(end)
  xval= linspace(xdata(1),xdata(end),NTEST);

  % we need the results of func at the points xdata to do the
  % interpolation WARNING: this is a vector statement
  % In a real problem, ydata would be "given" somehow, and
  % a function would not be available
  ydata=func(xdata);

  % use Lagrange interpolation from Chapter 5
  % WARNING: these use componentwise (vector) statements.
```

```
% Generate yval as interpolated values corresponding to xval
yval=eval_plin(xdata,ydata,xval);

% comparing yval with the exact results of func at xval
% In a real problem, the exact results would not be available.
yexact= func(xval);

% plot the exact and interpolated results on the same plot
% this gives assurance that everything is reasonable
plot(xval,yval,xval,yexact)

% compute the error in a standard way.
max_error=max(abs(yexact-yval))/max(abs(yexact));
end
```

(2) Include your plot here:

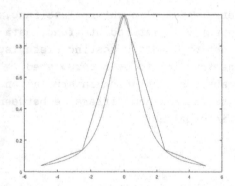

(3) Fill in the following table:

Runge function, Piecewise linear, uniformly-spaced points	
ndata	Maximum Error
5	0.18023
11	0.067442
21	0.041834
41	0.014040
81	0.0038013
161	9.6851e-04
321	2.4334e-04
641	6.0912e-05

(4) Repeat convergence using Chebyshev points:

Runge function, Piecewise linear, Chebyshev points	
ndata	Maximum Error
5	0.23552
11	0.059502
21	0.061458
41	0.028713
81	0.0088068
161	0.0023392
321	5.9581e-04
641	1.4991e-04

Solution 6.13.

(1) Include your plot here:

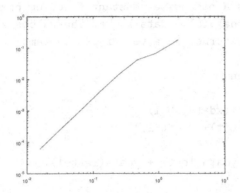

(2) Include your plot with slope estimates here:

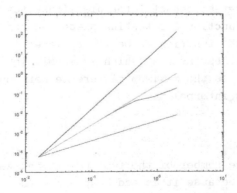

$C_1 = 3.9045\text{e-}3$, $C_2 = 2.5028\text{e-}1$, $C_3 = 1.6043\text{e}1$
The best slope estimate is p=2.

(3) List your error ratios: Max Error(5)/Max Error(11) =2.6724
Max Error(11)/Max Error(21) =1.6121
Max Error(21)/Max Error(41) =2.9797
Max Error(41)/Max Error(81) =3.6934
Max Error(81)/Max Error(161) =3.9248
Max Error(161)/Max Error(321)=3.9800
Max Error(320)/Max Error(641)=3.9950
Do they appear to approach 2^p for some integer p? (yes/no)
If so, what is your estimate of the value of p? $p = 2$

Solution 6.14.

(1) Include your eval_pconst.m here:

```
function yval = eval_pconst ( xdata, ydata, xval)
% yval = eval_pconst ( xdata, ydaa, xval)
% evaluates a piecewise constant function passing through the
% points (xdata(k), ydata(k)) at the values xval
% yval is returned as a vector of the same shape as xval

% M. Sussman

left=bracket(xdata,xval);
% y=.5*(y(left)+y(left+1))

yval= .5*( ydata(left) + ydata(left+1));
end
```

(2) Include your test_pconst_interpolate.m here:

```
function max_error=test_pconst_interpolate(func,xdata )
% max_error=test_pconst_interpolate(func,xdata )
% utility function for testing piecewise constant interpolation
% func is the function to be interpolated
% xdata are abscissae at which interpolation takes place
% max_error is the maximum difference between the function
%   and its interpolant

% M. Sussman

% Choose the number of the test points and generate them
% Use 4001 because it is odd, capturing the interval midpoint
NTEST=4001;
% construct NTEST points evenly spaced so that
% they cover the interpolation interval in a standard way, i.e.,
```

```
% xval(1)=xdata(1) and xval(NTEST)=xdata(end)
xval= linspace(xdata(1),xdata(end),NTEST);

% we need the results of func at the points xdata to do the
% interpolation WARNING: this is a vector statement
% In a real problem, ydata would be "given" somehow, and
% a function would not be available
ydata=func(xdata);

% use Lagrange interpolation from Chapter 5
% WARNING: these use componentwise (vector) statements.
% Generate yval as interpolated values corresponding to xval
yval=eval_pconst(xdata,ydata,xval);

% comparing yval with the exact results of func at xval
% In a real problem, the exact results would not be available.
yexact= func(xval);

% plot the exact and interpolated results on the same plot
% this gives assurance that everything is reasonable
plot(xval,yval,xval,yexact)

% compute the error in a standard way.
max_error=max(abs(yexact-yval))/max(abs(yexact));
end
```

(3) Include your plot here:

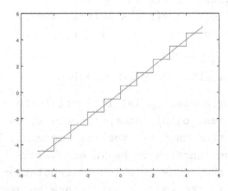

(4) Perform a convergence study by filling the following table using the Runge example function.

Runge function, Piecewise constant	
ndata	Maximum Error
5	0.4310345
11	0.2500000
21	0.1500000
41	0.0800000
81	0.0404494
161	0.0202714
321	0.0101469
641	0.0050736

(5) Estimated convergence rate p=1.

Solution 6.15.

(1) Include your modified `eval_plin.m`:

```
function yval = eval_plin ( xdata, ydata, xval)
  % yval = eval_plin ( xdata, ydaa, xval)
  % evaluates a piecewise linear function passing through the
  % points (xdata(k), ydata(k)) at the values xval
  % yval is returned as a vector of the same shape as xval

  % M. Sussman

  left=bracket(xdata,xval);
  % y=(y_b*(x-a) + y_a*(b-x))/(b-a)
  % is linear and =y_b at a and y_a at b

  yval=(ydata(left+1).*(xval-xdata(left)  ) + ...
        ydata(left  ).*(xdata(left+1)-xval)) ./  ...
        (xdata(left+1) - xdata(left));
end
```

(2) Include your `test_plin1_interpolate.m` here:

```
function max_error=test_plin1_interpolate(func,xdata )
  % max_error=test_plin1_interpolate(func,xdata )
  % utility function used for testing piecewise linear interpolation
  % func is the function to be interpolated
  % xdata are abscissae at which interpolation takes place
  % max_error is the maximum difference between the function
  %    and its interpolant

  % M. Sussman
```

```
% Choose the number of the test points and generate them
% Use 4001 because it is odd, capturing the interval midpoint
NTEST=4001;
% construct NTEST points evenly spaced so that
% they cover the interpolation interval in a standard way, i.e.,
% xval(1)=xdata(1) and xval(NTEST)=xdata(end)
xval= linspace(xdata(1),xdata(end),NTEST);

% we need the results of func at the points xdata to do the
% interpolation WARNING: this is a vector statement
% In a real problem, ydata would be "given" somehow, and
% a function would not be available
ydata=func(xdata);

% use Lagrange interpolation from Chapter 5
% WARNING: these use componentwise (vector) statements.
% Generate yval as interpolated values corresponding to xval
[yval,y1val]=eval_plin1(xdata,ydata,xval);

% comparing yval with the exact results of func at xval
% In a real problem, the exact results would not be available.
[yexact,y1exact]= func(xval);

% plot the exact and interpolated results on the same plot
% this gives assurance that everything is reasonable
plot(xval,y1val,xval,y1exact)

% compute the error in a standard way.
max_error=max(abs(y1exact-y1val))/max(abs(y1exact));
end
```

(3) Explain why you believe your result is correct.

The derivative of abs(x) is sign(x). The plot of the derivative is a multiple of the Heaviside function.

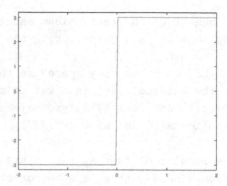

(4) Estimate the rate of convergence of the derivative. Explain your work. I also modified **runge.m** to return the derivative:

```
function [y,y1]=runge(x)
  % [y,y1]=runge(x)
  % computes the Runge example function and its derivative

  % M. Sussman

  y=1./(1+x.^2);
  y1=-(2*x)./(1+x.^2).^2;
end
```

Derivative of Runge function		
ndata	Maximum Error in derivative	ratio
5	0.530897	0.68966
11	0.769800	1.25000
21	0.615840	1.70000
41	0.362259	1.91176
81	0.189489	1.97692
161	0.095851	1.99416
321	0.048066	1.99854
641	0.024050	—

"Ratio" means the error on this line divided by the error on the following line. Rate of convergence of derivative, p=1.

Chapter 7

Higher-order interpolation

Solution 7.1. Include your plot here:

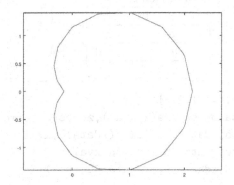

Solution 7.2.

(1) Include your `eval_pherm.m` here:

```
function yval = eval_pherm ( xdata, ydata, ypdata, xval)
  % yval = eval_plin ( xdata, ydata, ypdata,  xval)
  % evaluates a piecewise Hermite function passing through the
  % points (xdata(k), ydata(k)) at the values xval
  %   with derivative values at xdata from ypdata
  % yval is returned as a vector of the same shape as xval

  % M. Sussman

  k=bracket(xdata,xval);

  % 4 Hermite interpolation polynomials
  h1=(xval-xdata(k+1)).^2.*(3*xdata(k)-xdata(k+1)-2*xval)./ ...
                          (xdata(k)-xdata(k+1)).^3;
```

```
    h3=(xval-xdata(k)).^2.*(3*xdata(k+1)-xdata(k)-2*xval)./ ...
                    (xdata(k+1)-xdata(k)).^3;
    h2=(xval-xdata(k)).*(xval-xdata(k+1)).^2./ ...
                    (xdata(k)-xdata(k+1)).^2;
    h4=(xval-xdata(k+1)).*(xval-xdata(k)).^2./ ...
                    (xdata(k+1)-xdata(k)).^2;

    yval=ydata(k).*h1+ypdata(k).*h2+ydata(k+1).*h3+ypdata(k+1).*h4;
end
```

(2) Fill in the following table:

eval_pherm interpolated values				
your points →	0	1	2	3
function ↓				
$y = 1$	1	1	1	1
$y = x$	0	1	2	3
$y = x^2$	0	1	4	9
$y = x^3$	0	1	8	27

```
xdata=[0,2];xval=[0,1,2,3];
eval_pherm(xdata,ones(size(xdata)),zeros(size(xdata)),xval)
eval_pherm(xdata,xdata,ones(size(xdata)),xval)
eval_pherm(xdata,xdata.^2,2*xdata,xval)
eval_pherm(xdata,xdata.^3,3*xdata.^2,xval)
```

Solution 7.3.

(1) Include your modified runge.m here:

```
function [y,y1]=runge(x)
  % [y,y1]=runge(x)
  % computes the Runge example function and its derivative
  % y is the function, y1 is the derivative

  % M. Sussman

  y=1./(1+x.^2);
  y1=-(2*x)./(1+x.^2).^2;
end
```

(2) Include your test_pherm_interpolate.m here:

```
function max_error=test_pherm_interpolate(func,xdata )
  % max_error=test_pherm_interpolate(func,xdata )
  % utility function used testing piecewise Hermite interpolation
```

```
% func is the function to be interpolated
% xdata are abscissae at which interpolation takes place
% max_error is the maximum difference between the function
%   and its interpolant

% M. Sussman

% Choose the number of the test points and generate them
% Use 4001 because it is odd, capturing the interval midpoint
NTEST=4001;
% construct NTEST points evenly spaced so that
% they cover the interpolation interval in a standard way, i.e.,
% xval(1)=xdata(1) and xval(NTEST)=xdata(end)
xval= linspace(xdata(1),xdata(end),NTEST);

% we need the results of func at the points xdata to do the
% interpolation WARNING: this is a vector statement
% In a real problem, ydata would be "given" somehow, and
% a function would not be available
[ydata,ypdata]=func(xdata);

% use eval_pherm to do the interpolation
% WARNING: these use componentwise (vector) statements.
% Generate yval as interpolated values corresponding to xval
yval=eval_pherm(xdata,ydata,ypdata,xval);

% we will be comparing yval with the exact results of func at xval
% In a real problem, the exact results would not be available.
yexact= func(xval);

% plot the exact and interpolated results on the same plot
% this gives assurance that everything is reasonable
plot(xval,yval,xval,yexact)

% compute the error in a standard way.
max_error=max(abs(yexact-yval))/max(abs(yexact));
end
```

(3) Fill in the following table:

Runge function, Piecewise Hermite Cubic		
ndata	error	ratio
5	3.5621e-02	—
11	1.2522e-03	28.4463
21	1.8655e-04	6.7125
41	1.4275e-05	13.0686
81	9.3796e-07	15.2191
161	5.9171e-08	15.8516
321	3.7096e-09	15.9510
641	2.3203e-10	15.9877

(4) Estimate the rate of convergence: 4 $(15.9877 \approx 2^4)$

Solution 7.4.

(1) Include your exer4.m here:

```
% generate the data points for a limacon
numIntervals = 20;
tdata = linspace ( 0, 2*pi, numIntervals + 1 );
r = 1.15 + cos ( tdata );
xdata = r .* cos ( tdata );
xpdata = - r .* sin ( tdata ) - sin( tdata ) .* cos ( tdata );
ydata = r .* sin ( tdata );
ypdata = r .* cos( tdata) - sin(tdata).^2;

% interpolate them
tval = linspace ( 0, 2*pi, 10*(numIntervals+1) );
xval = eval_pherm ( tdata, xdata, xpdata, tval );
yval = eval_pherm ( tdata, ydata, ypdata, tval );

% plot them with correct aspect ratio
plot ( xval, yval )
axis equal
```

(2) Include your plot here:

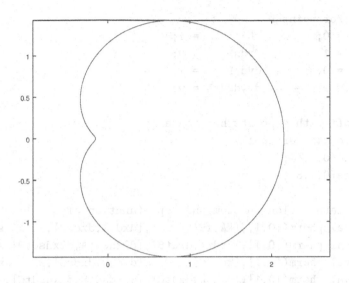

Solution 7.5.
Include your exer5.m here:

```
% Chapter 7, exercise 5
% File named exer5.m
% M. Sussman

% Generate a mesh based on Hermite bicubic interpolation

% values for point A
xA        = 0;          yA        = 0;
dxdsA     = 1;          dydsA     = 0;
dxdtA     = 0;          dydtA     = 1;
d2xdsdtA = 0;           d2ydsdtA = 0;

% values for point B
xB        = 1;          yB        = 0;
dxdsB     = 1;          dydsB     = 0;
dxdtB     = 0;          dydtB     = 1;
d2xdsdtB = 0;           d2ydsdtB = 0;

% values for point C
xC        = 1;          yC        = 1;
dxdsC     = 1;          dydsC     = 0;
dxdtC     = 0;          dydtC     = 1;
```

```
d2xdsdtC = 0;          d2ydsdtC = 0;

% values for point D
xD       = 0;          yD       = 1;
dxdsD    = 1;          dydsD    = 0;
dxdtD    = 0;          dydtD    = 1;
d2xdsdtD = 0;          d2ydsdtD = 0;

% Start off with 4 points horizontally,
% and 5 points vertically
s=linspace(0,1,4);
t=linspace(0,1,5);

% interpolate x along bottom and top, function of s
xAB    =eval_pherm([0,1],[xA,xB]     ,[dxdsA,dxdsB],       s);
dxdtAB=eval_pherm([0,1],[dxdtA,dxdtB],[d2xdsdtA,d2xdsdtB],s);
xDC    =eval_pherm([0,1],[xD,xC]     ,[dxdsD,dxdsC],       s);
dxdtDC=eval_pherm([0,1],[dxdtD,dxdtC],[d2xdsdtD,d2xdsdtC],s);

% interpolate y along bottom and top, function of s
yAB    =eval_pherm([0,1],[yA,yB]     ,[dydsA,dydsB],       s);
dydtAB=eval_pherm([0,1],[dydtA,dydtB],[d2ydsdtA,d2ydsdtB],s);
yDC    =eval_pherm([0,1],[yD,yC]     ,[dydsD,dydsC],       s);
dydtDC=eval_pherm([0,1],[dydtD,dydtC],[d2ydsdtD,d2ydsdtC],s);

% interpolate s-interpolations in t-direction
% if variables x and y already exist, they might have
% the wrong dimensions.  Get rid of them before reusing them.
clear x y
for k=1:length(s)
  x(k,:)=eval_pherm([0,1],[xAB(k),xDC(k)],[dxdtAB(k),dxdtDC(k)],t);
  y(k,:)=eval_pherm([0,1],[yAB(k),yDC(k)],[dydtAB(k),dydtDC(k)],t);
end

% plot all lines
plot(x(:,1),y(:,1),'b')
hold on
for k=2:numel(t)
  plot(x(:,k),y(:,k),'b')
end
for k=1:numel(s)
  plot(x(k,:),y(k,:),'b')
```

```
end
axis('equal');
hold off
```

Include your plot here:

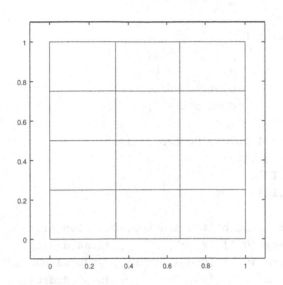

Solution 7.6.
Include your exer6.m here:

```
% Chapter 7, exercise 6
% File named exer6.m
% M. Sussman

% Generate a mesh based on Hermite bicubic interpolation

% values for point A
xA        = 0;          yA        = 0;
dxdsA     = 3;          dydsA     = 0;
dxdtA     = 0;          dydtA     = 1;
d2xdsdtA = 0;           d2ydsdtA = 0;

% values for point B
xB        = 1;          yB        = 0;
dxdsB     = .1;         dydsB     = 0;
dxdtB     = 0;          dydtB     = 1;
d2xdsdtB = 0;           d2ydsdtB = 0;
```

```
% values for point C
xC        = 1;          yC        = 1;
dxdsC     = .1;         dydsC     = 0;
dxdtC     = 0;          dydtC     = 1;
d2xdsdtC = 0;           d2ydsdtC = 0;


% values for point D
xD        = 0;          yD        = 1;
dxdsD     = 3;          dydsD     = 0;
dxdtD     = 0;          dydtD     = 1;
d2xdsdtD = 0;           d2ydsdtD = 0;


% Start off with 4 points horizontally,
% and 5 points vertically
s=linspace(0,1,11);
t=linspace(0,1,16);


% interpolate x along bottom and top, function of s
xAB   =eval_pherm([0,1],[xA,xB]      ,[dxdsA,dxdsB],        s);
dxdtAB=eval_pherm([0,1],[dxdtA,dxdtB],[d2xdsdtA,d2xdsdtB],s);
xDC   =eval_pherm([0,1],[xD,xC]      ,[dxdsD,dxdsC],        s);
dxdtDC=eval_pherm([0,1],[dxdtD,dxdtC],[d2xdsdtD,d2xdsdtC],s);


% interpolate y along bottom and top, function of s
yAB   =eval_pherm([0,1],[yA,yB]      ,[dydsA,dydsB],        s);
dydtAB=eval_pherm([0,1],[dydtA,dydtB],[d2ydsdtA,d2ydsdtB],s);
yDC   =eval_pherm([0,1],[yD,yC]      ,[dydsD,dydsC],        s);
dydtDC=eval_pherm([0,1],[dydtD,dydtC],[d2ydsdtD,d2ydsdtC],s);


% interpolate s-interpolations in t-direction
% if variables x and y already exist, they might have
% the wrong dimensions.  Get rid of them before reusing them.
clear x y
for k=1:length(s)
  x(k,:)=eval_pherm([0,1],[xAB(k),xDC(k)],[dxdtAB(k),dxdtDC(k)],t);
  y(k,:)=eval_pherm([0,1],[yAB(k),yDC(k)],[dydtAB(k),dydtDC(k)],t);
end


% plot all lines
plot(x(:,1),y(:,1),'b')
hold on
```

```
for k=2:numel(t)
  plot(x(:,k),y(:,k),'b')
end
for k=1:numel(s)
  plot(x(k,:),y(k,:),'b')
end
axis('equal');
hold off
```

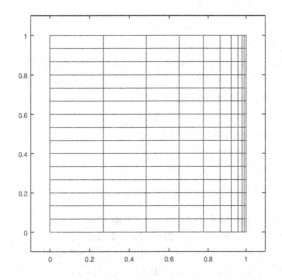

Solution 7.7.
Include your `exer7.m` here:

```
% Chapter 7, exercise 7
% File named exer7.m
% M. Sussman


% Generate a mesh based on Hermite bicubic interpolation

% values for point A
xA        = -1;        yA        = 0;
dxdsA     = 2;         dydsA     = 0;
dxdtA     = 1;         dydtA     = 1;
d2xdsdtA = -1;         d2ydsdtA = -0.5;


% values for point B
xB        = 1;         yB        = 0;
```

```
dxdsB     = 2;        dydsB     = 0;
dxdtB     = 0;        dydtB     = 0.5;
d2xdsdtB = -1;        d2ydsdtB = -0.5;


% values for point C
xC        = 1;        yC        = 0.5;
dxdsC     = 1;        dydsC     = -0.5;
dxdtC     = 0;        dydtC     = 0.5;
d2xdsdtC = -1;        d2ydsdtC = -0.5;


% values for point D
xD        = 0;        yD        = 1;
dxdsD     = 1;        dydsD     = -0.5;
dxdtD     = 1;        dydtD     = 1;
d2xdsdtD = -1;        d2ydsdtD = -0.5;


% Start off with 4 points horizontally,
% and 5 points vertically
s=linspace(0,1,20);
t=linspace(0,1,15);


% interpolate x along bottom and top, function of s
xAB    =eval_pherm([0,1],[xA,xB]         ,[dxdsA,dxdsB],        s);
dxdtAB=eval_pherm([0,1],[dxdtA,dxdtB],[d2xdsdtA,d2xdsdtB],s);
xDC    =eval_pherm([0,1],[xD,xC]         ,[dxdsD,dxdsC],        s);
dxdtDC=eval_pherm([0,1],[dxdtD,dxdtC],[d2xdsdtD,d2xdsdtC],s);


% interpolate y along bottom and top, function of s
yAB    =eval_pherm([0,1],[yA,yB]         ,[dydsA,dydsB],        s);
dydtAB=eval_pherm([0,1],[dydtA,dydtB],[d2ydsdtA,d2ydsdtB],s);
yDC    =eval_pherm([0,1],[yD,yC]         ,[dydsD,dydsC],        s);
dydtDC=eval_pherm([0,1],[dydtD,dydtC],[d2ydsdtD,d2ydsdtC],s);


% interpolate s-interpolations in t-direction
% if variables x and y already exist, they might have
% the wrong dimensions.  Get rid of them before reusing them.
clear x y
for k=1:length(s)
  x(k,:)=eval_pherm([0,1],[xAB(k),xDC(k)],[dxdtAB(k),dxdtDC(k)],t);
  y(k,:)=eval_pherm([0,1],[yAB(k),yDC(k)],[dydtAB(k),dydtDC(k)],t);
end
```

```
% plot all lines
plot(x(:,1),y(:,1),'b')
hold on
for k=2:numel(t)
  plot(x(:,k),y(:,k),'b')
end
for k=1:numel(s)
  plot(x(k,:),y(k,:),'b')
end
axis('equal');
hold off
```

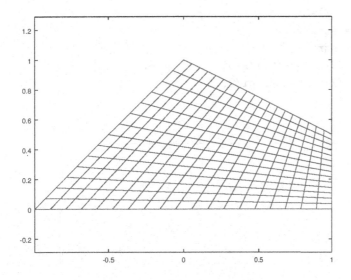

The mapping function turns out to be bilinear:

```
x=(1-t)*(2*s-1)+t*s
y=t*(1-0.5*s)
```

Solution 7.8.

(2) Include your exer8a.m here:

```
% Chapter 7, exercise 8a
% File named exer8a.m
% M. Sussman

% generate the data points for a limacon
numIntervals = 20;
s = linspace(0,1,numIntervals+1);
```

```
theta = 3*pi/4- s*pi/2;
r=sqrt(2)/2;
x = r .* cos ( theta )+.5;
y = r .* sin ( theta )-.5;

% plot them with correct aspect ratio
plot ( x, y )
axis equal

dxdt=-r*sin(theta);
dydt=r*cos(theta);
dtds=-pi/2;
% at s=0, theta(1)=3*pi/4;
% sin(3*pi/4)=-sqrt(2)/2
% cos(3*pi/4)= sqrt(2)/2
dxds= -sqrt(2)/2*(-sqrt(2))/2*pi/2;   % pi/4
dyds=  sqrt(2)/2*sqrt(2)/2*pi/2; %  pi/4
```

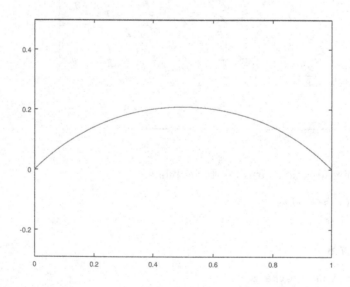

$$\frac{dx}{ds}\big|_A = \pi/4 \quad \frac{dy}{ds}\big|_A = \pi/4 \quad \frac{dx}{ds}\big|_B = \pi/4 \quad \frac{dy}{ds}\big|_B = -\pi/4$$

(3) Include your `exer8b.m` here:

```
% Chapter 7, exercise 8b
% File named exer8b.m
% M. Sussman
```

```
% Generate a mesh based on Hermite bicubic interpolation

% values for point A
xA       = 0;          yA        = 0;
dxdsA    = pi/4;          dydsA    = pi/4;

% values for point B
xB       = 1;          yB        = 0;
dxdsB    = pi/4;          dydsB    = -pi/4;

s=linspace(0,1,25);

% interpolate x along bottom and top, function of s
xAB    =eval_pherm([0,1],[xA,xB]        ,[dxdsA,dxdsB],        s);

% interpolate y along bottom and top, function of s
yAB    =eval_pherm([0,1],[yA,yB]        ,[dydsA,dydsB],        s);

% plot bottom (AB) line
plot(xAB,yAB,'r*-')
```

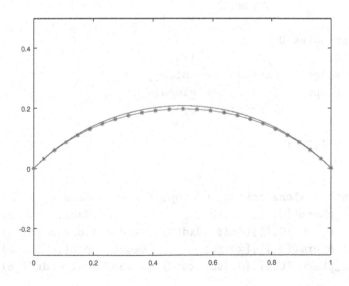

(4) Include your **exer8.m** here:

```
% Chapter 7,. exercise 8
% File named exer8.m
```

```
% M. Sussman

% Generate a mesh based on Hermite bicubic interpolation

slope=pi/4;

% values for point A
xA        = 0;        yA        = 0;
dxdsA     = slope;    dydsA     = slope;
dxdtA     = -slope;   dydtA     = slope;
d2xdsdtA = 0;         d2ydsdtA = 0;

% values for point B
xB        = 1;        yB        = 0;
dxdsB     = slope;    dydsB     = -slope;
dxdtB     = slope;    dydtB     = slope;
d2xdsdtB = 0;         d2ydsdtB = 0;

% values for point C
xC        = 1;        yC        = 1;
dxdsC     = slope;    dydsC     = slope;
dxdtC     = -slope;   dydtC     = slope;
d2xdsdtC = 0;         d2ydsdtC = 0;

% values for point D
xD        = 0;        yD        = 1;
dxdsD     = slope;    dydsD     = -slope;
dxdtD     = slope;    dydtD     = slope;
d2xdsdtD = 0;         d2ydsdtD = 0;

s=linspace(0,1,25);
t=linspace(0,1,25);

% interpolate x along bottom and top, function of s
xAB    =eval_pherm([0,1],[xA,xB]        ,[dxdsA,dxdsB],       s);
dxdtAB=eval_pherm([0,1],[dxdtA,dxdtB],[d2xdsdtA,d2xdsdtB],s);
xDC    =eval_pherm([0,1],[xD,xC]        ,[dxdsD,dxdsC],       s);
dxdtDC=eval_pherm([0,1],[dxdtD,dxdtC],[d2xdsdtD,d2xdsdtC],s);

% interpolate y along bottom and top, function of s
yAB    =eval_pherm([0,1],[yA,yB]        ,[dydsA,dydsB],       s);
dydtAB=eval_pherm([0,1],[dydtA,dydtB],[d2ydsdtA,d2ydsdtB],s);
```

```
yDC   =eval_pherm([0,1],[yD,yC]           ,[dydsD,dydsC],          s);
dydtDC=eval_pherm([0,1],[dydtD,dydtC],[d2ydsdtD,d2ydsdtC],s);

% interpolate s-interpolations in t-direction
% if variables x and y already exist, they might have
% the wrong dimensions.  Get rid of them before reusing them.
clear x y
for k=1:length(s)
  x(k,:)=eval_pherm([0,1],[xAB(k),xDC(k)],[dxdtAB(k),dxdtDC(k)],t);
  y(k,:)=eval_pherm([0,1],[yAB(k),yDC(k)],[dydtAB(k),dydtDC(k)],t);
end

% plot all lines
plot(x(:,1),y(:,1),'b')
hold on
for k=2:numel(t)
  plot(x(:,k),y(:,k),'b')
end
for k=1:numel(s)
  plot(x(k,:),y(k,:),'b')
end
axis('equal');
hold off
```

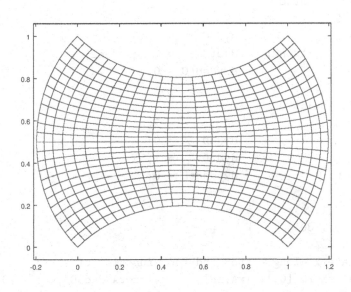

Solution 7.9.

(1) Include your exer9a.m here:

```
% Chapter 7, exercise 9
% File named exer9a.txt
% M. Sussman

% Generate a mesh for the patch ABCD
sqrt2on2=sqrt(2)/2;

% values for point A
xA        = 0;             yA        = 1;
dxdsA     = pi/4;          dydsA     = 0;
dxdtA     = 0;             dydtA     = 1;
d2xdsdtA = 0;             d2ydsdtA = 0;

% values for point B
xB        = sqrt2on2;      yB        = sqrt2on2;
dxdsB     = sqrt2on2*pi/4; dydsB     = -sqrt2on2*pi/4;
dxdtB     = 2-sqrt2on2;    dydtB     = 2-sqrt2on2;
d2xdsdtB = 0;             d2ydsdtB = 0;

% values for point C
xC        = 2;             yC        = 2;
dxdsC     = 2;             dydsC     = 0;
dxdtC     = 2-sqrt2on2;    dydtC     = 2-sqrt2on2;
d2xdsdtC = 0;             d2ydsdtC = 0;

% values for point D
xD        = 0;             yD        = 2;
dxdsD     = 2;             dydsD     = 0;
dxdtD     = 0;             dydtD     = 1;
d2xdsdtD = 0;             d2ydsdtD = 0;

s=linspace(0,1,20);
t=linspace(0,1,30);

% interpolate x along bottom and top, function of s
xAB    =eval_pherm([0,1],[xA,xB]        ,[dxdsA,dxdsB],        s);
dxdtAB=eval_pherm([0,1],[dxdtA,dxdtB],[d2xdsdtA,d2xdsdtB],s);
xDC    =eval_pherm([0,1],[xD,xC]        ,[dxdsD,dxdsC],        s);
dxdtDC=eval_pherm([0,1],[dxdtD,dxdtC],[d2xdsdtD,d2xdsdtC],s);
```

```
% interpolate y along bottom and top, function of s
yAB   =eval_pherm([0,1],[yA,yB]       ,[dydsA,dydsB],       s);
dydtAB=eval_pherm([0,1],[dydtA,dydtB],[d2ydsdtA,d2ydsdtB],s);
yDC   =eval_pherm([0,1],[yD,yC]       ,[dydsD,dydsC],       s);
dydtDC=eval_pherm([0,1],[dydtD,dydtC],[d2ydsdtD,d2ydsdtC],s);

% interpolate s-interpolations in t-direction
% if variables x and y already exist, they might have
% the wrong dimensions.  Get rid of them before reusing them.
clear x y
for k=1:length(s)
  x(k,:)=eval_pherm([0,1],[xAB(k),xDC(k)],[dxdtAB(k),dxdtDC(k)],t);
  y(k,:)=eval_pherm([0,1],[yAB(k),yDC(k)],[dydtAB(k),dydtDC(k)],t);
end

% plot all lines
plot(x(:,1),y(:,1),'b')
hold on
for k=2:numel(t)
  plot(x(:,k),y(:,k),'b')
end
for k=1:numel(s)
  plot(x(k,:),y(k,:),'b')
end
axis('equal');
hold off
```

(2) Include your `exer9b.m` here:

```
% Chapter 7, exercise 9
% File named exer9b.txt
% M. Sussman

% Generate a mesh for the patch ABCD
sqrt2on2=sqrt(2)/2;

% values for point A
xA       = 0;              yA       = 1;
dxdsA    = pi/4;           dydsA    = 0;
dxdtA    = 0;              dydtA    = 1;
d2xdsdtA = 0;             d2ydsdtA = 0;
```

```
% values for point B
xB       = sqrt2on2;        yB       = sqrt2on2;
dxdsB    = sqrt2on2*pi/4;  dydsB    = -sqrt2on2*pi/4;
dxdtB    = 2-sqrt2on2;      dydtB    = 2-sqrt2on2;
d2xdsdtB = 0;               d2ydsdtB = 0;

% values for point C
xC       = 2;               yC       = 2;
dxdsC    = 2;               dydsC    = 0;
dxdtC    = 2-sqrt2on2;      dydtC    = 2-sqrt2on2;
d2xdsdtC = 0;               d2ydsdtC = 0;

% values for point D
xD       = 0;               yD       = 2;
dxdsD    = 2;               dydsD    = 0;
dxdtD    = 0;               dydtD    = 1;
d2xdsdtD = 0;               d2ydsdtD = 0;

s=linspace(0,1,20);
t=linspace(0,1,30);

% interpolate x along bottom and top, function of s
xAB   =eval_pherm([0,1],[xA,xB]       ,[dxdsA,dxdsB],       s);
dxdtAB=eval_pherm([0,1],[dxdtA,dxdtB],[d2xdsdtA,d2xdsdtB],s);
xDC   =eval_pherm([0,1],[xD,xC]       ,[dxdsD,dxdsC],       s);
dxdtDC=eval_pherm([0,1],[dxdtD,dxdtC],[d2xdsdtD,d2xdsdtC],s);

% interpolate y along bottom and top, function of s
yAB   =eval_pherm([0,1],[yA,yB]       ,[dydsA,dydsB],       s);
dydtAB=eval_pherm([0,1],[dydtA,dydtB],[d2ydsdtA,d2ydsdtB],s);
yDC   =eval_pherm([0,1],[yD,yC]       ,[dydsD,dydsC],       s);
dydtDC=eval_pherm([0,1],[dydtD,dydtC],[d2ydsdtD,d2ydsdtC],s);

% interpolate s-interpolations in t-direction
clear x y
for k=1:length(s)
  x(k,:)=eval_pherm([0,1],[xAB(k),xDC(k)],[dxdtAB(k),dxdtDC(k)],t);
  y(k,:)=eval_pherm([0,1],[yAB(k),yDC(k)],[dydtAB(k),dydtDC(k)],t);
end

% plot all lines
plot(x(:,1),y(:,1),'b')
```

```
hold on
for k=2:numel(t)
  plot(x(:,k),y(:,k),'b')
end
for k=1:numel(s)
  plot(x(k,:),y(k,:),'b')
end
axis('equal');

% values for point a
xa       = 1;            ya       = 0;
dxdsa    = 1;            dydsa    = 0;
dxdta    = 0;            dydta    = pi/4;
d2xdsdta = 0;           d2ydsdta = 0;

% values for point b
xb       = 2;            yb       = 0;
dxdsb    = 1;            dydsb    = 0;
dxdtb    = 0;            dydtb    = 2;
d2xdsdtb = 0;           d2ydsdtb = 0;

% values for point c
xc       = 2;            yc       = 2;
dxdsc    = 2-sqrt2on2;   dydsc    = 2-sqrt2on2;
dxdtc    = 0;            dydtc    = 2;
d2xdsdtc = 0;           d2ydsdtc = 0;

% values for point d
xd       = sqrt2on2;     yd       = sqrt2on2;
dxdsd    = 2-sqrt2on2;   dydsd    = 2-sqrt2on2;
dxdtd    = -sqrt2on2*pi/4;dydtd    = sqrt2on2*pi/4;
d2xdsdtd = 0;           d2ydsdtd = 0;

s=linspace(0,1,30);
t=linspace(0,1,20);

% interpolate x along bottom and top, function of s
xab    =eval_pherm([0,1],[xa,xb]         ,[dxdsa,dxdsb],      s);
dxdtab=eval_pherm([0,1],[dxdta,dxdtb],[d2xdsdta,d2xdsdtb],s);
xdc    =eval_pherm([0,1],[xd,xc]         ,[dxdsd,dxdsc],      s);
dxdtdc=eval_pherm([0,1],[dxdtd,dxdtc],[d2xdsdtd,d2xdsdtc],s);
```

```
% interpolate y along bottom and top, function of s
yab   =eval_pherm([0,1],[ya,yb]         ,[dydsa,dydsb],       s);
dydtab=eval_pherm([0,1],[dydta,dydtb],[d2ydsdta,d2ydsdtb],s);
ydc   =eval_pherm([0,1],[yd,yc]         ,[dydsd,dydsc],       s);
dydtdc=eval_pherm([0,1],[dydtd,dydtc],[d2ydsdtd,d2ydsdtc],s);

% interpolate s-interpolations in t-direction
clear x y
for k=1:length(s)
  x(k,:)=eval_pherm([0,1],[xab(k),xdc(k)],[dxdtab(k),dxdtdc(k)],t);
  y(k,:)=eval_pherm([0,1],[yab(k),ydc(k)],[dydtab(k),dydtdc(k)],t);
end

% plot all lines
plot(x(:,1),y(:,1),'b')
hold on
for k=2:numel(t)
  plot(x(:,k),y(:,k),'b')
end
for k=1:numel(s)
  plot(x(k,:),y(k,:),'b')
end
axis('equal');
hold off
```

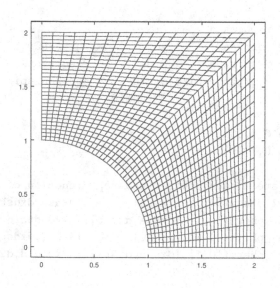

Solution 7.10.

(1) Include your ccspline.m here:

```
function sprime=ccspline(xdata,ydata,y1p,ynp)
  % sprime=ccsspline(xdata,ydata,y1p,ynp)
  % complete cubic spline
  % (xdata,ydata) are points through which the spline passes
  % y1p is the slope at the first endpoint
  % ynp is the slope at the final endpoint
  % sprime is the values of the slopes at the knots

  % M. Sussman

  n=numel(xdata);

  % (7.7)
  for k=1:n-1
    h(k)=xdata(k+1)-xdata(k);
  end

  % (7.12)
  D(1,1)= (ydata(2)-ydata(1))/h(1) - y1p;
  for k=2:n-1
    D(k,1)= (ydata(k+1)-ydata(k))/h(k) ...
            -(ydata(k)-ydata(k-1))/h(k-1);
  end
  D(n,1)= ynp - (ydata(n)-ydata(n-1))/h(n-1);

  % (7.11)
  A=zeros(n,n);
  A(1,1)= h(1)/3;
  A(1,2)= h(1)/6;
  for k=2:n-1
    A(k,k-1)= h(k-1)/6;
    A(k,k)   = (h(k-1)+h(k))/3;
    A(k,k+1)= h(k)/6;
  end
  A(n,n-1)= h(n-1)/6;
  A(n,n)   = h(n-1)/3;

  M=A\D;
```

```
% (7.8)
sprime=zeros(size(ydata));
sprime(1)=y1p;
for k=2:n-1
   sprime(k)=-M(k)*h(k)/3-M(k+1)*h(k)/6 ...
                 +(ydata(k+1)-ydata(k))/h(k);
end
sprime(n)=ynp;
end
```

(2) What are your results for (sprime)? [0 3 12 48]
    Are they correct? (yes/no)

Solution 7.11.

(1) Include your test_ccspline_interpolate.m here:

```
function max_error=test_ccspline_interpolate(func,xdata )
  % max_error=test_ccspline_interpolate(func,xdata )
  % utility function used for testing cubic spline interpolation
  % func is the function to be interpolated
  % xdata are abscissae at which interpolation takes place
  % max_error is the maximum difference between the function
  %    and its interpolant

  % M. Sussman

  % Choose the number of the test points and generate them
  % Use 4001 because it is odd, capturing the interval midpoint
  NTEST=4001;
  % construct NTEST points evenly spaced so that
  % they cover the interpolation interval in a standard way, i.e.,
  % xval(1)=xdata(1) and xval(NTEST)=xdata(end)
  xval= linspace(xdata(1),xdata(end),NTEST);

  % we need the results of func at the points xdata to do the
  % interpolation WARNING: this is a vector statement
  % In a real problem, ydata would be "given" somehow, and
  % a function would not be available
  [ydata,ypdata]=func(xdata);

  % use eval_pherm to do the interpolation
  % WARNING: these use componentwise (vector) statements.
  % Generate yval as interpolated values corresponding to xval
```

```
ypdata=ccspline(xdata,ydata,ypdata(1),ypdata(end));
yval=eval_pherm(xdata,ydata,ypdata,xval);

% we will be comparing yval with the exact results of func at xval
% In a real problem, the exact results would not be available.
yexact= func(xval);

% plot the exact and interpolated results on the same plot
% this gives assurance that everything is reasonable
plot(xval,yval,xval,yexact)

% compute the error in a standard way.
max_error=max(abs(yexact-yval))/max(abs(yexact));
end
```

(2) Fill in the following table.

Runge function, Complete Cubic Spline		
ndata	error	ratio
5	5.6067e-02	—
11	3.1829e-03	17.615
21	2.7797e-04	11.450
41	1.6107e-05	17.257
81	9.6744e-07	16.649
161	5.9647e-08	16.220
321	3.7166e-09	16.049
641	2.3214e-10	16.010

(3) The estimated rate of convergence is 4.

Solution 7.12.

(1) Include your ncspline.m here:

```
function sprime=ncspline(xdata,ydata)
  % sprime=ncsspline(xdata,ydata)
  % natural cubic spline
  % (xdata,ydata) are points through which the spline passes
  % sprime is the values of the slopes at the knots

  % M. Sussman

  n=numel(xdata);

  % (7.7)
```

```
    for k=1:n-1
      h(k)=xdata(k+1)-xdata(k);
    end


    % (7.12)
    D(1,1)= 0; % slope=0 at ends for natural splines
    for k=2:n-1
      D(k,1)= (ydata(k+1)-ydata(k))/h(k) ...
                -(ydata(k)-ydata(k-1))/h(k-1);
    end
    D(n,1)= 0; % slope=0 at ends for natural splines


    % (7.11)
    A=zeros(n,n);
    A(1,1)= h(1)/3;
    for k=2:n-1
      A(k,k-1)= h(k-1)/6;
      A(k,k)   = (h(k-1)+h(k))/3;
      A(k,k+1)= h(k)/6;
    end
    A(n,n)   = h(n-1)/3;


    M=A\D;


    % (7.8)
    sprime=zeros(size(ydata));
    sprime(1)=-M(1)*h(1)/3-M(2)*h(1)/6+(ydata(2)-ydata(1))/h(1);
    for k=2:n-1
      sprime(k)=-M(k)*h(k)/3-M(k+1)*h(k)/6 ...
                    +(ydata(k+1)-ydata(k))/h(k);
    end
    sprime(n)=M(n)*h(n-1)/3+M(n-1)*h(n-1)/6 ...
                    +(ydata(n)-ydata(n-1))/h(n-1);
  end
```

(2) Does your interpolate agree with the linear function? (yes/no)
(3) Does M(1)=0 and M(end)=0? (yes/no)
(4) Do ncsprime and ccsprime agree up to roundoff?
    (yes/no)
(5) Include your test_ncspline_interpolate.m here:

```
function max_error=test_ncspline_interpolate(func,xdata )
  % max_error=test_ncspline_interpolate(func,xdata )
```

```
% utility for testing natural cubic spline interpolation
% func is the function to be interpolated
% xdata are abscissae at which interpolation takes place
% max_error is the maximum difference between the function
%   and its interpolant

% M. Sussman

% Choose the number of the test points and generate them
% Use 4001 because it is odd, capturing the interval midpoint
NTEST=4001;
% construct NTEST points evenly spaced so that
% they cover the interpolation interval in a standard way, i.e.,
% xval(1)=xdata(1) and xval(NTEST)=xdata(end)
xval= linspace(xdata(1),xdata(end),NTEST);

% we need the results of func at the points xdata to do the
% interpolation WARNING: this is a vector statement
% In a real problem, ydata would be "given" somehow, and
% a function would not be available
ydata=func(xdata);

% use eval_pherm to do the interpolation
% WARNING: these use componentwise (vector) statements.
% Generate yval as interpolated values corresponding to xval
ypdata=ncspline(xdata,ydata);
yval=eval_pherm(xdata,ydata,ypdata,xval);

% comparing yval with the exact results of func at xval
% In a real problem, the exact results would not be available.
yexact= func(xval);

% plot the exact and interpolated results on the same plot
% this gives assurance that everything is reasonable
plot(xval,yval,xval,yexact)

% compute the error in a standard way.
max_error=max(abs(yexact-yval))/max(abs(yexact));
end
```

(6) Fill in the following table:

Runge function, Natural Cubic Spline		
ndata	error	ratio
5	8.3312e-02	—
11	3.0422e-02	2.7385
21	6.7604e-03	4.5001
41	1.5725e-03	4.2992
81	3.8589e-04	4.0749
161	9.5842e-05	4.0264
321	2.3933e-05	4.0047
641	5.8722e-06	4.0756

(7) The estimated rate of convergence is 2.

Solution 7.13.

(1) Include your test_spline_interpolate.m here:

```
function max_error=test_spline_interpolate(func,xdata )
  % max_error=test_spline_interpolate(func,xdata )
  % utility for testing Matlab spline interpolation
  % func is the function to be interpolated
  % xdata are abscissae at which interpolation takes place
  % max_error is the maximum difference between the function
  %    and its interpolant

  % M. Sussman

  % Choose the number of the test points and generate them
  % Use 4001 because it is odd, capturing the interval midpoint
  NTEST=4001;
  % construct NTEST points evenly spaced so that
  % they cover the interpolation interval in a standard way, i.e.,
  % xval(1)=xdata(1) and xval(NTEST)=xdata(end)
  xval= linspace(xdata(1),xdata(end),NTEST);

  % we need the results of func at the points xdata to do the
  % interpolation WARNING: this is a vector statement
  % In a real problem, ydata would be "given" somehow, and
  % a function would not be available
  ydata=func(xdata);

  % Generate yval from Matlab spline function
```

```
yval=spline(xdata,ydata,xval);

% comparing yval with the exact results of func at xval
% In a real problem, the exact results would not be available.
yexact= func(xval);

% plot the exact and interpolated results on the same plot
% this gives assurance that everything is reasonable
plot(xval,yval,xval,yexact)

% compute the error in a standard way.
max_error=max(abs(yexact-yval))/max(abs(yexact));
end
```

(2) Fill in the following table:

Runge function, Not-a-knot Cubic Spline		
ndata	error	ratio
5	1.0845e-01	—
11	1.2389e-02	8.7533
21	6.5990e-04	18.7741
41	1.2614e-04	5.2316
81	9.6978e-06	13.0070
161	6.3643e-07	15.2376
321	4.0000e-08	15.9108
641	2.4961e-09	16.0250

(3) The estimated rate of convergence is 4.

Solution 7.14.

(3) Include your plot of all three curves here:

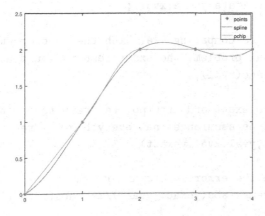

Solution 7.15.

(1) Include your test_pchip_interpolate.m here:

```
function max_error=test_pchip_interpolate(func,xdata )
  % max_error=test_spline_interpolate(func,xdata )
  % utility function used for testing Matlab monotone interpolation
  % func is the function to be interpolated
  % xdata are abscissae at which interpolation takes place
  % max_error is the maximum difference between the function
  %   and its interpolant

  % M. Sussman

  % Choose the number of the test points and generate them
  % Use 4001 because it is odd, capturing the interval midpoint
  NTEST=4001;
  % construct NTEST points evenly spaced so that
  % they cover the interpolation interval in a standard way, i.e.,
  % xval(1)=xdata(1) and xval(NTEST)=xdata(end)
  xval= linspace(xdata(1),xdata(end),NTEST);

  % we need the results of func at the points xdata to do the
  % interpolation WARNING: this is a vector statement
  % In a real problem, ydata would be "given" somehow, and
  % a function would not be available
  ydata=func(xdata);

  % Generate yval from Matlab spline function
  yval=pchip(xdata,ydata,xval);

  % we will be comparing yval with the exact results of func at xval
  % In a real problem, the exact results would not be available.
  yexact= func(xval);

  % plot the exact and interpolated results on the same plot
  % this gives assurance that everything is reasonable
  plot(xval,yval,xval,yexact)

  % compute the error in a standard way.
  max_error=max(abs(yexact-yval))/max(abs(yexact));
end
```

(2) Fill in the following table:

Runge function, Monotone Cubic Interpolation		
ndata	error	ratio
5	1.0191e-01	—
11	3.0616e-02	3.3286
21	5.9143e-03	5.1766
41	1.2417e-03	4.7632
81	2.9462e-04	4.2144
161	7.2551e-05	4.0609
321	1.8024e-05	4.0252
641	4.5015e-06	4.0040

(3) The estimated rate of convergence is 2.

Chapter 8

Legendre polynomials and L^2 approximation

Solution 8.1.

(1) Include a copy of your coef_mon.m here:

```
function c=coef_mon(func,n)
  % c=coef_mon(func,n) coefficients of monomial L2 approximation
  % func is a function handle
  % n is the number of monomials, highest power=(n-1)

  % M. Sussman

  for k=1:n
    % force b to be a column vector with second index
    b(k,1)=ntgr8(@(x) func(x).*x.^(n-k));
  end

  for k=1:n
    for ell=1:n
      H(k,ell)=ntgr8(@(x) x.^(2*n-k-ell));
    end
  end

  c=H\b;

end
```

(2) What coefficients did you get? 3, -2, 1
    Are they correct? (<u>yes</u>/no)

(3) Include a copy of your test_mon.m here:

```
% Chapter 8, exercise 1
% File named test_mon.m
```

```
% M. Sussman

func=@partly_quadratic;
c=coef_mon(func,n);

xval=linspace(-1,1,10000);
yval=polyval(c,xval);
yexact=func(xval);
plot(xval,yval,xval,yexact)

% relative Euclidean norm is approximating
% the relative integral least-squares (L2 norm)
% using an approximate trapezoid rule
relativeError=norm(yexact-yval)/norm(yexact)
```

Is the error for n=5 smaller than for n=1? (yes/no)
Include your plot for the case n=5 here:

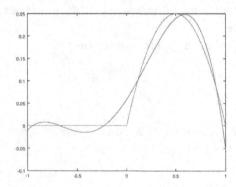

Does it appear that the area between the exact and approximate curves is divided equally between "above" and "below?" (yes/no)

(4) Fill in the following table using the Runge example function.

Runge test function	
n	relative error
1	2.0057e-01
2	2.0057e-01
3	3.5030e-02
4	3.5030e-02
5	6.0506e-03
6	6.0506e-03
10	1.7915e-04
20	2.6769e-08
30	1.5689e-08
40	4.8770e-09

(5) Explain why the errors in the Runge case for **n=1** and **n=2**, *etc.*, are the same. Runge is an even function. Integrating and even function times an odd function such as x^3 gives zero, so the odd coefficients are zero and the corresponding even and odd approximations are the same.

(6) Fill in the following table for the partly quadratic function.

partly_quadratic function	
n	relative error
1	0.76379
2	0.52055
3	0.51550
4	0.28872
5	0.15732
6	0.14237
10	0.061659
20	0.020791
30	0.018003
40	0.015352

(7) Fill in the following table for the sawshape8 function.

sawshape8 function	
n	relative error
1	1
2	0.86607
3	0.86607
4	0.64955
5	0.64955
6	0.54129
10	0.42627
20	0.30520
30	0.29161
40	0.29754

Solution 8.2.

(1) Include your copy of `legen.m` here:

```
function yval=legen(n,xval)
  % yval=legen(n,xval) computes the Legendre function P_n

  % M. Sussman

  if n<0
    error('legen: n must be nonnegative.')
  elseif n==0
    yval=ones(size(xval));
  elseif n==1;
    yval=xval;
  else
    ykm1=ones(size(xval));
    yk=xval;
    for k=2:n
      ykm2=ykm1;
      ykm1=yk;
      yk=((2*k-1)*xval.*ykm1-(k-1)*ykm2)/k;
    end
    yval=yk;
  end
end
```

(2) Execute the given code. What is the printed value? 3.0179e-16
    Is it of roundoff size? (<u>yes</u>/no)
(3) Execute the given code. What is the printed value? 0
    Is it of roundoff size? (<u>yes</u>/no)

Solution 8.3.

(1) Include your coef_legen.m here:

```
function d=coef_legen(func,n)
  % d=coef_legen(func,n)
  % d= coefficients of the Legendre L2 approximation of func

  % M. Sussman

  for k=1:n
    d(k)=(2*k-1)/2*ntgr8(@(x) func(x).*legen(k-1,x));
  end
end
```

(2) What values for d did you get?
2.4286e-17 -7.2858e-17 1.0408e-16 1.0000e+00
Are they correct? (<u>yes</u>/no)

(3) Include your eval_legen.m here:

```
function yval=eval_legen(d,xval)
  % yval=eval_legen(d,xval)
  % evaluate the Legendre expansion y=sum(d_k*legen(xval,k))

  % M. Sussman

  n=numel(d);
  yval=d(1)*legen(0,xval);
  for k=2:n
    yval=yval+d(k)*legen(k-1,xval);
  end
end
```

(4) Do your five values of eval_legen and legen agree? (<u>yes</u>/no)
0 1 17 63 154

(5) Include your test_legen.m here:

```
% File named test_legen.m
% M. Sussman

func=@runge;
%func=@partly_quadratic;
%func=@sawshape8;

tic;
```

```
d=coef_legen(func,n);

xval=linspace(-1,1,10000);
yval=eval_legen(d,xval);
yexact=func(xval);
plot(xval,yval,xval,yexact)

% relative Euclidean norm is approximating
% the relative integral least-squares (L2 norm)
% using an approximate trapezoid rule
relativeError=norm(yexact-yval)/norm(yexact)
toc;
```

(6) Nothing required
(7) Fill in the following table for the Runge example function.

Runge test function	
n	relative error
1	0.20057
2	0.20057
3	0.035030
4	0.035030
5	0.0060506
6	0.0060506
10	1.7915e-04
20	2.6768e-08
30	3.9886e-12
40	1.0683e-15
50	1.1562e-15

(8) Fill in the following table for the partly quadratic function. Include the elapsed time only for the final three lines.

partly_quadratic function		
n	relative error	elapsed time
5	0.15732	—
10	0.061659	—
20	0.020791	—
40	0.0071941	—
80	0.0025179	—
160	8.8586e-04	1.627310e+00
320	3.1234e-04	6.755532e+00
640	1.1013e-04	3.338850e+01

(9) Fill in the following table for the sawshape8 function. Include the elapsed time only for the final three lines.

sawshape8 function		
n	relative error	elapsed time
5	0.64955	—
10	0.42627	—
20	0.30520	—
40	0.21716	—
80	0.15404	—
160	0.10909	1.797729
320	0.077188	8.538394
640	0.054574	46.74888

(10) If elapsed time is proportional to n^p, $2 \le p \le 3$ because doubling the problem size increases the elapsed time by more than a factor of 4 but less than a factor of 8.

Solution 8.4.

(1) Include your coef_fourier.m here:

```
function [z,s,c]=coef_fourier(func,n)
  % [z,s,c]=coef_fourier(func,n)
  % z= coefficients of the constant in Fourier approx of func
  % s= coefficients of the sin in Fourier approx of func
  % c= coefficients of the cos in Fourier approx of func

  % M. Sussman

  z=1/sqrt(2)*ntgr8(func);

  for k=1:n
    s(k)=ntgr8(@(x) func(x).*sin(k*pi*x));
    c(k)=ntgr8(@(x) func(x).*cos(k*pi*x));
  end
end
```

(2) Did you get $z = 1$, and all others zero in the first case? (yes/no)
$s_2 = 1$ and all others zero in the second case? (yes/no)
$c_3 = 1$ with all others zero in the third case? (yes/no)

(3) Include your eval_fourier.m here:

```
function yval=eval_fourier(z,s,c,xval)
  % yval=eval_legen(z,s,c,xval)
```

```
% evaluate the Fourier expansion

% M. Sussman

n=numel(s);
if numel(z) ~= 1
  error('eval_fourier: z must be a scalar.');
end
if numel(c) ~= n
  error('eval_fourier: s and c must have the same size.');
end
yval=z/sqrt(2)*ones(size(xval));
for k=1:n
  yval=yval+s(k)*sin(k*pi*xval)+c(k)*cos(k*pi*xval);
end
end
```

(4) For $f(x) = 1/\sqrt{2}$, what values did you choose and what results did you get?

```
xval=linspace(-1,1,100);
[z,s,c]=coef_fourier(@(x) ones(size(x))/sqrt(2),3);
norm(eval_fourier(z,s,c,xval)-ones(size(xval))/sqrt(2))
ans =    2.6132e-15
```

For $f(x) = \sin(2\pi x)$, what values did you choose and what results did you get?

```
xval=linspace(-1,1,100);
[z,s,c]=coef_fourier(@(x) sin(2*pi*x),3);
norm(eval_fourier(z,s,c,xval)-sin(2*pi*xval))
ans =    2.0620e-15
```

For $f(x) = \cos(3\pi x)$, what values did you choose and what results did you get?

```
xval=linspace(-1,1,100);
[z,s,c]=coef_fourier(@(x) cos(3*pi*x),3);
norm(eval_fourier(z,s,c,xval)-cos(3*pi*xval))
ans =    2.3465e-15
```

(5) Include your test_fourier.m here:

```
% File named test_fourier.m
% M. Sussman

func=@runge;
```

```
%func=@partly_quadratic;
%func=@sawshape8;

tic;
[z,s,c]=coef_fourier(func,n);

xval=linspace(-1,1,10000);
yval=eval_fourier(z,s,c,xval);
yexact=func(xval);
plot(xval,yval,xval,yexact)

% relative Euclidean norm is approximating
% the relative integral least-squares (L2 norm)
% using an approximate trapezoid rule
relativeError=norm(yexact-yval)/norm(yexact)
toc
```

(6) Fill in the following table for the Runge example function.

Runge test function		
n	relative error	elapsed time
1	2.111119e-02	—
2	1.273385e-02	—
3	7.723479e-03	—
4	5.346594e-03	—
5	3.973877e-03	—
6	3.101218e-03	—
10	1.517809e-03	—
50	1.459561e-04	—
100	5.278086e-05	6.178190e-01
200	1.930365e-05	1.539305e+00
400	7.259940e-06	4.339447e+00
800	2.886306e-06	1.797448e+01

(7) Fill in the following table for the partly quadratic function.

partly_quadratic function		
n	relative error	elapsed time
1	2.899957e-01	—
2	8.406764e-02	—
3	7.989048e-02	—
4	3.957656e-02	—
5	3.917078e-02	—
6	2.413782e-02	—
10	1.235975e-02	—
50	1.253262e-03	—
100	4.531952e-04	5.653610e-01
200	1.638837e-04	1.429503e+00
400	5.993738e-05	3.453079e+00
800	2.254193e-05	9.147433e+00

(8) Fill in the following table for the sawshape8 function.

sawshape8 function		
n	relative error	elapsed time
1	6.261572e-01	—
2	4.899909e-01	—
3	4.153837e-01	—
4	3.668080e-01	—
5	3.320106e-01	—
6	3.055227e-01	—
10	2.405286e-01	—
50	1.097122e-01	—
100	7.776224e-02	6.408780e-01
200	5.502761e-02	1.608398e+00
400	3.885747e-02	4.464786e+00
800	2.726502e-02	1.449348e+01

(9) For your estimate of elapsed time, $1 \leq p \leq 2$

Solution 8.5.

(1) Include your coef_pc.m here:

```
function a=coef_pc(func,Npc)
  % a=coef_pc(func,Npc)
  % a=coefficients of piecewise constant approximation of func

  % M. Sussman
```

```
      Npt=Npc+1;   % one more point than interval
      xpt=linspace(-1,1,Npt);
      for k=1:Npc
         a(k)=Npc/2*integrate(func,xpt(k),xpt(k+1));
      end
   end
```

(2) Are all $a_k = 1$? (<u>yes</u>/no)
(3) Are your values correct? (<u>yes</u>/no)
     [-0.9 -0.7 -0.5 -0.3 -0.1 0.1 0.3 0.5 0.7 0.9]
(4) Include your eval_pc.m here:

```
   function yval=eval_pc(a,xval)
     % yval=coef_pc(a,xval)
     % a= coefficients of the piecewise constant approximation of func
     % xval=values for evaluation

     % M. Sussman

     Npc=numel(a);
     Npt=Npc+1;   % one more point than interval
     xpt=linspace(-1,1,Npt);
     xleft=bracket(xpt,xval);

     yval=a(xleft);
   end
```

(5) Are your answers correct? (<u>yes</u>/no)
(6) Include your test_pc.m here:

```
   % File named test_pc.m
   % M. Sussman

   func=@runge;
   %func=@partly_quadratic;
   %func=@sawshape8;

   tic;
   a=coef_pc(func,Npc);

   xval=linspace(-1,1,20000);
   yval=eval_pc(a,xval);
   yexact=func(xval);
   plot(xval,yval,xval,yexact)
```

```
% relative Euclidean norm is approximating
% the relative integral least-squares (L2 norm)
% using an approximate trapezoid rule
relativeError=norm(yexact-yval)/norm(yexact)
toc
```

(7) Do you believe that no other piecewise constant function would produce a better approximation? (yes/no)

(8) Fill in the following table for the Runge example function:

Runge test function		
n	relative error	elapsed time
4	9.429628e-02	—
8	4.747276e-02	—
16	2.378798e-02	—
64	5.951211e-03	—
256	1.487826e-03	—
1024	3.719577e-04	1.153855e+00
4096	9.298938e-05	4.546849e+00
16384	2.324735e-05	1.836725e+01

(9) Fill in the following table for the partly quadratic function:

partly_quadratic function		
n	relative error	elapsed time
4	4.082993e-01	—
8	2.224669e-01	—
16	1.134076e-01	—
64	2.851966e-02	—
256	7.133191e-03	—
1024	1.783350e-03	1.187435e+00
4096	4.458381e-04	4.682947e+00
16384	1.114595e-04	2.045056e+01

(10) Fill in the following table for the sawshape8 function:

	sawshape8 function		
n	relative error	elapsed time	
4	2.500188e-01	—	
8	1.250094e-01	—	
16	6.250469e-02	—	
64	1.562617e-02	—	
256	3.906543e-03	—	
1024	9.766357e-04	1.183882e+00	
4096	2.441589e-04	4.694706e+00	
16384	6.103973e-05	2.052822e+01	

(11) What is your estimate of the integer p where **relative error** is proportional to $(1/n)^p$? p=1

(12) What is your estimate of the integer p where **elapsed time** is proportional to n^p? p=1

Solution 8.6.

(1) Include your **hat.m** here:

```
function yval=hat(k,Np1,xval)
  % yval=hat(k,Np1,xval) defined in Section 8.9
  % k is index of this hat function
  % Np1 is the number of subintervals of [-1,1]
  % x is the location(s) at which to evaluate the function

  % M. Sussman

  x=linspace(-1,1,Np1+1);
  left=bracket(x,xval);

  yval=zeros(size(xval));
  if k<=Np1
    yval(left==k)=(x(k+1)-xval(left==k))/(x(k+1)-x(k));
  end
  if k>=2
    yval(left==k-1)=(xval(left==k-1)-x(k-1))/(x(k)-x(k-1));
  end
end
```

(2) Include your plot here:

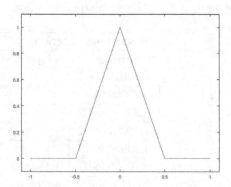

(3) Include your coef_plin.m here:

```
function a=coef_plin(func,Npl)
  % a=coef_plin(func,Npl)
  % a= coefficients of the piecewise linear approximation of func

  % M. Sussman

  Npt=Npl+1;   % one more point than interval
  xpt=linspace(-1,1,Npt);

  rhs(1)=integrate(@(x) hat(1,Npl,x).*func(x),xpt(1),xpt(2));
  for k=2:Npl
    rhs(k)=integrate( ...
            @(x) hat(k,Npl,x).*func(x),xpt(k-1),xpt(k+1));
  end
  rhs(Npl+1)=integrate( ...
            @(x) hat(Npl+1,Npl,x).*func(x),xpt(Npl),xpt(Npl+1));

  H=zeros(Npl+1,Npl+1);
  H(1,1)=2/(3*Npl);
  H(1,2)=1/(3*Npl);
  for k=2:Npl
    H(k,k)=4/(3*Npl);
    H(k,k+1)=1/(3*Npl);
    H(k,k-1)=1/(3*Npl);
  end
  H(Npl+1,Npl+1)=2/(3*Npl);
  H(Npl+1,Npl)=1/(3*Npl);

  a=H\rhs'; % rhs is a row vector
```

```
end
```

Tests:

```
coef_plin(@(x) hat(1,5,x),5) = [1; 0; 0; 0; 0; 0;]
coef_plin(@(x) hat(2,5,x),5) = [0; 1; 0; 0; 0; 0;]
coef_plin(@(x) hat(3,5,x),5) = [0; 0; 1; 0; 0; 0;]
coef_plin(@(x) hat(4,5,x),5) = [0; 0; 0; 1; 0; 0;]
coef_plin(@(x) hat(5,5,x),5) = [0; 0; 0; 0; 1; 0;]
coef_plin(@(x) hat(6,5,x),5) = [0; 0; 0; 0; 0; 1;]
```

(4) Include your `eval_plin.m` here:

```
function yval=eval_plin(c,xval)
  % yval=eval_plin(c,xval)
  % evaluate the piecewise linear expansion y=sum(c_k*hat(k,x))

  % M. Sussman

  Npl=numel(c)-1;
  yval=c(1)*hat(1,Npl,xval);
  for k=2:Npl+1
    yval=yval+c(k)*hat(k,Npl,xval);
  end
end
```

Test:

```
Npl=5;
c=coef_plin(@(x) x,Npl);
norm(eval_plin(c,linspace(-1,1,15))-linspace(-1,1,15)) = 0
```

(5) Include your `test_plin.m` here:

```
% File named test_plin.m
% M. Sussman

%func=@runge;
%func=@partly_quadratic;
func=@sawshape8;

tic;
a=coef_plin(func,Npc);

xval=linspace(-1,1,20000);
yval=eval_plin(a,xval);
```

```
yexact=func(xval);
plot(xval,yval,xval,yexact)

% relative Euclidean norm is approximating
% the relative integral least-squares (L2 norm)
% using an approximate trapezoid rule
relativeError=norm(yexact-yval)/norm(yexact)
toc
```

(6) Fill in the following table for the Runge example function:

Runge test function		
n	relative error	elapsed time
4	1.431037e-02	—
16	7.520718e-04	—
64	4.632137e-05	1.110770e-01
256	2.892431e-06	4.171490e-01
1024	1.807669e-07	1.660686e+00

- What is your estimate of the integer p where **relative error** is proportional to $(1/n)^p$? p=2
- What is your estimate of the integer p where **elapsed time** is proportional to n^p? p=1

(7) Fill in the following table for the partly quadratic function:

partly_quadratic function		
n	relative error	elapsed time
4	1.189082e-01	—
16	6.661154e-03	—
64	4.032001e-04	1.154290e-01
256	2.499076e-05	4.253280e-01
1024	1.558649e-06	1.686057

- What is your estimate of the integer p where **relative error** is proportional to $(1/n)^p$? p=2
- What is your estimate of the integer p where **elapsed time** is proportional to n^p? p=1

(8) Fill in the following table for the sawshape8 function:

	sawshape8 function	
n	relative error	elapsed time
4	6.546560e-01	—
16	3.290185e-01	—
64	1.645091e-01	1.156390e-01
256	8.225146e-02	4.237730e-01
1024	4.111814e-02	1.692968e+00

- What is your estimate of the integer p where **relative error** is proportional to $(1/n)^p$? p=1
- What is your estimate of the integer p where **elapsed time** is proportional to n^p? p=1

Chapter 9

Quadrature

Solution 9.1.

(1) Include a copy of your `midpointquad.m` here:

```
function quad = midpointquad( func, a, b, N)
  % quad = midpointquad( func, a, b, N)
  % func is a function handle for the function being integrated
  %    func must be able to accept a vector input and give a vector
  %    output
  % a and b are limits of integration
  % N is the number of points to use, including endpoints

  % M. Sussman

  xpts = linspace(a,b,N) ;
  h = xpts(2)-xpts(1) ; % length of subintervals
  xmidpts = 0.5 * ( xpts(1:N-1) + xpts(2:N) );
  fmidpts = func(xmidpts);
  quad = h * sum ( fmidpts );
end
```

(2) Did you get $\int_0^1 2x\,dx = 1$? (yes/no)

(3) Fill in the following table, using scientific notation for the error values so you can see the pattern.

Midpoint rule			
N	h	Midpoint result	Error
11	1.0	2.7363	0.010494
101	0.1	2.7363	1.2326e-05
1001	0.01	2.7363	1.2327e-07
10001	0.001	2.7363	1.2337e-09

(4) Your estimate of the order of accuracy is $O(h^2)$.

Solution 9.2.

(1) I use N=2 to fill in the following table:

func	Midpoint result	Error
1	1	0
$2x$	1	0
$3x^2$	.75	-.25
$4x^3$	.5	-.5

(2) What is the degree of exactness of the midpoint rule? 1
(3) Is the degree of exactness is one less than the order of accuracy? (yes,no)

Solution 9.3.

(1) Include your `trapezoidquad.m` here:

```
function quad = trapezoidquad( func, a, b, N)
% quad = trapezoidquad( func, a, b, N)
% func is a function handle for the function being integrated
%    func must be able to accept a vector input and give a vector
%    output
% a and b are limits of integration
% N is the number of points to use, including endpoints

% M. Sussman

xpts = linspace(a,b,N) ;
fmidpts = 0.5 * ( func(xpts(1:N-1)) + func(xpts(2:N)) );
h = xpts(2:N)-xpts(1:N-1);
quad = sum ( h.*fmidpts );
end
```

(2) Fill in the following table.

func	Trapezoid result	Error
1	1	0
$2x$	1	0
$3x^2$	1.5	0.5
$4x^3$	2	1

(3) What is the degree of exactness of the trapezoid rule? 1
(4) Fill in the following table, recording the error using scientific notation.

Runge example function			
N	h	Trapezoid result	Error
11	1.0	2.7561	9.3071e-03
101	0.1	2.7468	2.4653e-05
1001	0.01	2.7468	2.4655e-07
10001	0.001	2.7468	2.4655e-09

(5) What is the order of accuracy? $O(h^2)$

(6) Is the degree of exactness is one less than the order of accuracy? (yes/no)

Solution 9.4. Fill in the following table.

$\int_0^1 \log(x)\, dx$			
N	h	Midpoint result	Error
11	1.0	-0.96576	3.4241e-02
101	0.1	-0.99654	3.4616e-03
1001	0.01	-0.99965	3.4653e-04
10001	0.001	-0.99997	3.4657e-05

The estimated rate of convergence is $O(h^1)$.

Solution 9.5.

(1) Include a copy of your nc_single.m here:

```
function quad = nc_single ( func, a, b, N )
  % quad = nc_single ( func, a, b, N )
  % Newton-Cotes quadrature, single interval
  % func is a function handle for the integrand
  % a and b are the limits of the integral
  % N is the number of points to use

  % M. Sussman

  xvec = linspace ( a, b, N );
  wvec = nc_weight ( N );
  fvec = func(xvec);
  quad = (b-a) * sum(wvec .* fvec);

end
```

(2) For $N = 2$: $\int_0^1 2x\, dx = 1$ exactly? (yes/no)

(3) Fill in the following table using nc_single:

func	Error ($N = 4$)	Error ($N = 5$)	Error ($N = 6$)
$4x^3$	0	0	2.2204e-16
$5x^4$	1.8519e-02	0	2.2204e-16
$6x^5$	5.5556e-02	0	2.2204e-16
$7x^6$	1.0905e-01	2.6042e-03	1.4667e-03
Degree	3	5	5

Solution 9.6. Fill in the following table using `nc_single`

N	nc_single result	Error
3	6.7949	4.0481
7	3.8704	1.1236
11	4.6733	1.9265
15	7.8995	5.1527

Solution 9.7.

(1) Include a copy of your `nc_quad.m` here:

```
function quad = nc_quad ( func, a, b, N , numSubintervals)
  % quad = nc_quad ( func, a, b, N )
  % Newton-Cotes quadrature
  % func is a function handle for the integrand
  % a and b are the limits of the integral
  % N is the number of points to use
  % numSubintervals is the number of subintervals to use

  % M. Sussman

  x = linspace ( a, b, numSubintervals+1 );
  quad=0;
  for k=1:numSubintervals
    quad=quad+nc_single(func,x(k),x(k+1),N);
  end
end
```

(2) Which line did you choose? N=15
   What result did you get? quad=7.8995, error=5.1527
(3) Compute $\int_{-1}^{1} f_{\text{partly quadratic}}(x)\, dx$ using at least N=3 and `numSubintervals=2`
   quad=.16667
   Explain why your result should have an error of zero or roundoff-sized.
   With two intervals, the integral breaks into two polynomial integrands that are integrated exactly.

(4) Compute $\int_{-1}^{1} f_{\text{partly quadratic}}(x)\,dx$ using at least N=3 and numSubintervals=3
What result did you get? quad=1.4815e-01, error=-1.8519e-02 Explain why
your result should *not* have an error of zero or roundoff-sized.
The middle subinterval has a non-polynomial integrand

(5) Does your version get 2.74533025409679? (<u>yes</u>/no)

(6) Fill in the following table using the Runge function on [-5,5].

Subintervals	N	nc_quad error	ratio
10	2	9.3071e-03	1.5685e+01
20	2	-5.9337e-04	3.8525e+00
40	2	-1.5402e-04	3.9987e+00
80	2	-3.8519e-05	3.9997e+00
160	2	-9.6305e-06	3.9999e+00
320	2	-2.4077e-06	–
10	3	-3.8935e-03	5.1397e+02
20	3	-7.5753e-06	4.4327e+02
40	3	-1.7090e-08	1.6002e+01
80	3	-1.0680e-09	1.5995e+01
160	3	-6.6773e-11	1.5999e+01
320	3	-4.1736e-12	–
10	4	-1.4713e-03	5.1448e+02
20	4	-2.8597e-06	3.7652e+02
40	4	-7.5952e-09	1.6000e+01
80	4	-4.7470e-10	1.5995e+01
160	4	-2.9677e-11	1.5991e+01
320	4	-1.8558e-12	–

(7) For N=2, estimated order of convergence = 2
For N=3, estimated order of convergence = 4
For N=4, estimated order of convergence = 4

Solution 9.8.

(1) Include a copy of your gl_single.m here:

```
function quad = gl_single ( func, a, b, N )
  % quad = gl_single ( func, a, b, N )
  % Gauss-Legendre integration, single subinterval
  % func is a function handle for the integrand
  % a and b are the limits of the integral
  % N is the number of Gauss points to use

  % M. Sussman
```

```
    [xvec,wvec] = gl_weight (a, b, N );
    fvec = func(xvec);
    quad = sum(wvec .* fvec);

end
```

(2) Is its exactness is at least 1? (yes/no)

(3) Fill in the following table: The degree of exactness of the method is $2N - 1$,

func	Error ($N = 2$)	Error ($N = 3$)
$3x^2$	0	0
$4x^3$	0	0
$5x^4$	-2.7778e-02	0
$6x^5$	-8.3333e-02	2.2204e-16
$7x^6$	-1.5741e-01	-2.5000e-03
Degree	3	5

Is the degree of exactness of the method is $2N - 1$? (yes/no)

(4) Fill in the following table:

N	gl_single result	Error
3	4.7917	2.0449
7	3.0806	3.3381e-01
11	2.8123	6.5489e-02
15	2.7601	1.3266e-02

Solution 9.9.

(1) Include a copy of your gl_quad.m here:

```
function quad = gl_quad ( func, a, b, N , numSubintervals)
    % quad = gl_quad ( func, a, b, N ) Gauss-Legendre quadrature
    % func is a function handle for the integrand
    % a and b are the limits of the integral
    % N is the number of points to use
    % numSubintervals is the number of subintervals to use

    % M. Sussman

    x = linspace ( a, b, numSubintervals+1 );
    quad=0;
    for k=1:numSubintervals
       quad=quad+gl_single(func,x(k),x(k+1),N);
    end
end
```

(2) Which line did you choose? N=15

What result did you get? integral=2.7601e+00, error= 1.3266e-02

Do they agree? (yes/no)

(3) Compute $\int_{-1}^{1} f_{\text{partly quadratic}}(x)\, dx$ using numSubintervals=2 and $N \geq 2$.

integral=1.6667e-01, error=2.7756e-17

The integral is exact because with 2 intervals, the integrand is a continuous polynomial on each.

(4) Compute $\int_{-1}^{1} f_{\text{partly quadratic}}(x)\, dx$ using numSubintervals=3 and $N \geq 2$.

integral=1.7526e-01, error=8.5945e-03

The integral is not exact because with 3 intervals, the integrand is not a continuous polynomial on the middle interval.

(5) Do you get the specified result? (yes/no)

(6) Fill in the following table using the Runge function on [-5,5].

Subintervals	N	gl_quad error	ratio
10	1	-1.0494e-02	36.779
20	1	2.8532e-04	3.7061
40	1	7.6987e-05	3.9977
80	1	1.9258e-05	3.9994
160	1	4.8152e-06	3.9999
320	1	1.2038e-06	—
10	2	2.8067e-03	514.69
20	2	5.4532e-06	478.64
40	2	1.1393e-08	16.002
80	2	7.1198e-10	15.994
160	2	4.4516e-11	15.995
320	2	2.7831e-12	—
45	3	8.2823e-13	6.4310e+01
90	3	1.2879e-14	—
46	3	6.2084e-13	5.5920e+01
92	3	1.1102e-14	—
47	3	6.0218e-13	7.1368e+01
94	3	8.4377e-15	—
48	3	5.0049e-13	7.5133e+01
96	3	6.6613e-15	—
49	3	4.5830e-13	6.4500e+01
98	3	7.1054e-15	—

(7) Index 1, estimated order of convergence is 2

Index 2, estimated order of convergence is 4

(8) Index 3, estimated order of convergence is 6

Solution 9.10.

(1) Use gl_quad to fill in the following table.

$\int_0^1 \log(x)\, dx$			
Subintervals	N	gl_quad error	Error ratio
10	1	3.4241e-02	1.9879
20	1	1.7225e-02	1.9940
40	1	8.6383e-03	1.9970
80	1	4.3257e-03	—

What is the order of accuracy of the method using N=1? 1

(2) Use gl_quad to fill in the following table.

$\int_0^1 \log(x)\, dx$			
Subintervals	N	gl_quad error	Error ratio
10	2	1.0447e-02	2.0000
20	2	5.2237e-03	2.0000
40	2	2.6119e-03	2.0000
80	2	1.3059e-03	—

What is the order of accuracy of the method using N=2? 1

(3) Use gl_quad to fill in the following table.

$\int_0^1 \log(x)\, dx$			
Subintervals	N	gl_quad error	Error ratio
10	3	5.2334e-03	2.0000
20	3	2.6167e-03	2.0000
40	3	1.3083e-03	2.0000
80	3	6.5417e-04	—

What is the order of accuracy of the method using N=3? 1

Solution 9.11.

What integration method did you use? Gauß-Legendre, it is most efficient in terms of accuracy versus number of points and subintervals.

For N=1, the accuracy with 8 subintervals is 2.6039e-3. To see how many subintervals would be necessary to reach an accuracy of 1.e-8, with each doubling of the number of subintervals increasing the accuracy by a factor of 4, it would take

$$\log(2.6039 \cdot 10^{-3}/10^{-8})/\log(4) = 9$$

or 9 doublings of 8 subintervals, or 4096 subintervals: too many!

For N=2, the accuracy with 8 subintervals is 5.3734e-8. From there, 11 subintervals will bring it below 1.e-8 (trial and error). One doubling (16 subintervals) is predicted to bring the error down to 3.3584e-9, well below 1.e-8.

Solution 9.12.

(1) Include a copy of your adaptquad.m here:

```
function [Q,errEst,x,recursions]= ...
          adaptquad(func,x0,x1,tol,recursions)
% [Q,errEst,x,recursions]=
%          adaptquad(func,x0,x1,tol,recursions)
% adaptive quadrature
%     input parameters
% func         = function to integrate
% x0           = left end point
% x1           = right end point
% tol          = desired accuracy
% recursions = number of allowable recursions left
%
%     output parameters
% Q            = estimate of the value of the integral
% errEst       = estimate of error in Q
% x            = all intermediate integration points
% recursions = minimum number of recursions remaining
%                   after convergence

% M. Sussman


% Add a mid-point and re-estimate integral
xmid=(x0+x1)/2;

% Qleft and Qright are integrals over two halves
N=3;
Qboth=gl_single(func,x0,x1,N);
Qleft=gl_single(func,x0,xmid,N);
Qright=gl_single(func,xmid,x1,N);

% p=degree of exactness of Gauss-Legendre
p=2*N-1;
errEst=abs(Qleft+Qright-Qboth)/(2^(p+1)-1);

if errEst<tol | recursions<=0  %vertical bar means "or"
   % either ran out of recursions or converged
```

```
      Q= Qleft+Qright;
      x=[x0 xmid x1];
    else
      % not converged -- do it again
      [Qleft,estLeft,xleft,recursLeft]=adaptquad(func, ...
                    x0,xmid,tol/2,recursions-1);
      [Qright,estRight,xright,recursRight]=adaptquad(func, ...
                    xmid,x1,tol/2,recursions-1);
      % recursive work is all done, return answers
      % don't want xmid to appear twice in x
      x=[xleft xright(2:length(xright))];
      Q= Qleft+Qright;
      errEst= estLeft+estRight ;
      recursions=min(recursLeft,recursRight);
    end
  end
```

(2) What is the result of the integration?q=1.00000
    What is recursions? 5
    What is x? [0,.5,1.0]

(3) What is the result of the integration?q=1.00000
    What is recursions? 4
    What is x? [0,0.25,0.50,0.75,1.00]
    What are both the estimated (6.1035e-07) and true (1-q=6.1035e-07) errors?
(4) Fill in the following table

adaptquad for the Runge function		
tol	error estimate	exact error
1.e-3	1.3925e-04	7.8364e-04
1.e-6	2.7097e-07	2.5302e-07
1.e-9	3.0629e-10	2.1771e-10

Solution 9.13.
What are the final values of x?
x=[0,0.25,0.50,0.625,0.75,0.875,1.0]
What is the final value of recursions? 48

Solution 9.14.

(1) Include a copy of your srunge.m

```
    function y=srunge(x)
      % y=srunge(x)=1/(a^2+x^2) with a=1.e-3.
```

```
% M. Sussman

a=1.e-3;
y=1./(a^2+x.^2);
```

(2) What is the estimated error? 2.2939e-11
    What is the true error? 2.3192e-11
    Is recursions larger than zero? (yes/no)
(3) How many subintervals did gl_quad require? 9400
    How does this compare with the number of subintervals that adaptquad used?
    1552
(4) Include a copy of your plot here:

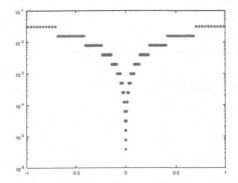

(5) What are the lengths of the largest and smallest intervals? 0.031250, 3.8147e-06
    Explain where to expect the smallest intervals for an arbitrary function.
    near a singularity or where the derivative is very large

Solution 9.15.

(1) What is the approximate value? q=1.4604
    What is the estimated error? 2.5660e-11
    What is the true error? q-sqrt(2)/6-sqrt(6)/2 = 2.7374e-11
(2) What is the returned value of recursions? 11

(3) Include your plot here:

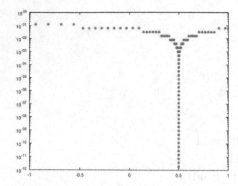

Solution 9.16.

(1) What is the computed value of $\int_0^1 x^{-0.99} dx = 100$? 32.311
    What is the estimated error? 0.0074733
    What is the true error? 100-q=67.689
    What is the value returned for **recursions**? 0
(2) For **recursions=60**: What is the estimated error? 0.0069728
    What is the true error? 63.156
    What is the value returned for **recursions**? 0

Chapter 10

Topics in quadrature and roundoff

Solution 10.1.

(1) The estimated error is 0.001634
and the true error is 0.0017807?.
Again: the estimated error is 0.001634
and the true error is 0.0014527.
These values come from pseudorandom numbers. Your answers will probably
be different.
Explain why both $\langle f \rangle$ and $\langle f^2 \rangle$ are the same value.
Since f in this case is ϕ, the characteristic function of a ball, it takes on the
values 0 and 1 only. f and f^2 are the same.

(2) The estimated error is 0.00094304
and the true error is 0.0026363.
These values come from pseudorandom numbers.
How many trials did you use? 1600*10000= 16e6. (Took original number of
trials, got estimated error of about 0.004, so multiplied the number of trials by
16.)

Solution 10.2.

(1) What is the computed value of $\int_0^2 e^x dx$? 6.3897
What is the estimated error? 8.9376e-4
How many trials did you use? 16,000,000
What is the true error? 6.6888e-4
Include your code here:

```
% Chapter 10, exercise 2.1
% File named exer2_1.m
% M. Sussman
% compute \int_0^2 e^x\,dx

CHUNK=10000;  % chosen for efficiency
```

```
NUM_CHUNKS=1600;

VOLUME=2;   % outer interval is [0,2]

totalPoints=0;
averagef=0;
averagef2=0;
for k=1:NUM_CHUNKS
  x=2*rand(CHUNK,1); % inside interval [0,2]
  f=exp(x);
  averagef=averagef+sum(f);
  averagef2=averagef2+sum(f.^2);
  totalPoints=totalPoints+CHUNK;
end
averagef=averagef/totalPoints;
averagef2=averagef2/totalPoints;
q=VOLUME*averagef;
disp(strcat('approx integral=',num2str(q), ...
              ' with true error=',num2str(exp(2)-1-q), ...
              ' and estimated error=', ...
      num2str(VOLUME*sqrt((averagef2-averagef^2)/totalPoints))));
```

(2) The computed value of $\int_\Omega e^{(x^2+y^2)} \, dxdy$ is 5.3976
What is the estimated error? is 3.9e-4
How many trials did you use? 20,000,000
What is the true error? 5.436e-4
Include your code here:

```
% Chapter 10, exercise 2.2
% File named exer12_2.m
% M. Sussman
% compute \int_{\Omega} e^(x^2+y^2)\,dxdy, \Omega=R^2 unit ball

CHUNK=10000;  % chosen for efficiency
NUM_CHUNKS=2000;

VOLUME=pi;  % volume of unit ball = pi

totalPoints=0;
averagef=0;
averagef2=0;
for k=1:NUM_CHUNKS
  x=(2*rand(CHUNK,1)-1);
```

```
    y=(2*rand(CHUNK,1)-1);
    phi=(x.^2+y.^2)<=1;
    f=exp(x.^2+y.^2).*phi; % only use x and y inside the ball
    averagef=averagef+sum(f);
    averagef2=averagef2+sum(f.^2);
    totalPoints=totalPoints+sum(phi); %only count points inside ball
  end
  averagef=averagef/totalPoints;
  averagef2=averagef2/totalPoints;
  q=VOLUME*averagef;
  disp(strcat('approx integral=',num2str(q), ...
              ' with true error=',num2str((exp(1)-1)*pi-q), ...
              ' and estimated error=', ...
         num2str(VOLUME*sqrt((averagef2-averagef^2)/totalPoints))));
```

(3) What is the computed value of the integral? 7.5934
     What is the estimated error? 1.8596e-3
     How many trials did you use? 20,000,000
     Include your code here:

```
% Chapter 10, exercise 2.3
% File named exer12_3.m
% M. Sussman
% compute \int_{\Omega} e^(x+y+z)dxdydz, \Omega=Steinmetz solid

CHUNK=10000;  % chosen for efficiency
NUM_CHUNKS=2000;

VOLUME=16/3;  % volume of Steinmetz solid

totalPoints=0;
averagef=0;
averagef2=0;
for k=1:NUM_CHUNKS
  x=(2*rand(CHUNK,1)-1);
  y=(2*rand(CHUNK,1)-1);
  z=(2*rand(CHUNK,1)-1);
  phi=( (x.^2+y.^2) <= 1 ) .* ( (x.^2 + z.^2) <= 1 );
  f=exp(x+y+z).*phi; %only use x, y inside the Steinmetz solid
  averagef=averagef+sum(f);
  averagef2=averagef2+sum(f.^2);
  totalPoints=totalPoints+sum(phi); %only count points inside
end
```

```
averagef=averagef/totalPoints;
averagef2=averagef2/totalPoints;
q=VOLUME*averagef;
disp(strcat('approx integral=',num2str(q), ...
        ' with estimated error=', ...
    num2str(VOLUME*sqrt((averagef2-averagef^2)/totalPoints))));
```

Solution 10.3.

(1) Include you q_elt.m here:

```
function q=q_elt(f,x,y,h)
  % q=q_elt(f,x,y,h)
  % f=function handle for function to be integrated
  % x=lower left point abcissa
  % y=lower left point ordinate
  % h=length of side of (square) element
  % q=integral of f over this element

  % M. Sussman

  s=1/(2*sqrt(3));
  X(1)=x+h*(1/2+s);
  X(2)=x+h*(1/2+s);
  X(3)=x+h*(1/2-s);
  X(4)=x+h*(1/2-s);
  Y(1)=y+h*(1/2+s);
  Y(2)=y+h*(1/2-s);
  Y(3)=y+h*(1/2+s);
  Y(4)=y+h*(1/2-s);

  q=h^2/4*sum(f(X,Y));

end
```

(2) Test q_elt on the functions 1, $4xy$, $6x^2y$, $9x^2y^2$, and $16x^3y^3$ over the square $[0,1] \times [0,1]$ and show that the result is exact, up to roundoff.

Function	Integral	Error
1	1	0
$4xy$	1	0
$6x^2y$	1	0
$9x^2y^2$	1	0
$16x^3y^3$	1	0

(3) Test q_elt on the function $25x^4y^4$ to see that it is not exact, thus showing the degree of precision is 3.

Integral = 0.94522

Error = 0.054784

Solution 10.4.

(1) Include your qerr_elt.m here:

```
function [q,errest]=qerr_elt(f,x,y,h)
  % [q,errest]=qerr_elt(f,x,y,h)
  % f=function handle for function to be integrated
  % x=lower left point abcissa
  % y=lower left point ordinate
  % h=length of side of (square) element
  % q=integral of f over this element
  % errest= estimate from Equation (10.10)

  % M. Sussman

  qh=q_elt(f,x,y,h);

  q1=q_elt(f,x      ,y      ,h/2);
  q2=q_elt(f,x+h/2,y      ,h/2);
  q3=q_elt(f,x+h/2,y+h/2,h/2);
  q4=q_elt(f,x      ,y+h/2,h/2);

  q=q1+q2+q3+q4;
  errest=(qh-q1-q2-q3-q4)/15;
end
```

(2) $\int_0^1 \int_0^1 16x^3y^3 \, dxdy = 1$

Error estimate = 0

True error = 0

(3) $\int_0^1 \int_0^1 25x^4y^4 \, dxdy = 0.99653$

Error estimate =0.0034210

True error =0.0034692

Is the estimated error within 5% of the true error? (<u>yes</u>/no)

Solution 10.5.

(2) Partial q_total=1

(3) Second test of partial q_total=4

(4) Include your final version of q_total.m here:

```
function [q,errest]=q_total(f,x,y,H,n)
```

```
% [q,errest]=q_total(f,x,y,H,n)
% n=number of intervals along one side
% f=function handle for function to be integrated
% x=lower left point abcissa
% y=lower left point ordinate
% H=length of side
% q=integral of f
% errest= error estimate

% M. Sussman

h=(H)/n;
eltCount=0;
for k=1:n
  for j=1:n
    eltCount=eltCount+1;
    elt(eltCount).x= x+(k-1)*h;
    elt(eltCount).y= y+(j-1)*h;
    elt(eltCount).h= h;
    [elt(eltCount).q,elt(eltCount).errest] = ...
       qerr_elt(f,elt(eltCount).x,elt(eltCount).y,h);
  end
end
if numel(elt) ~= n^2
  error('q_total: wrong number of elements!')
end

q=0;
errest=0;
for k=1:numel(elt);
  q=q+elt(k).q;
  errest=errest+abs(elt(k).errest);
end
end
```

(5) $\int_0^1 \int_0^1 9x^2y^2 \, dxdy = 1$
Estimated error $= 0$
True error $= 0$

(6) $\int_{-1}^1 \int_{-1}^1 9x^2y^2 \, dxdy = 4$
Estimated error $= 4$
True error $= 5.9212\mathrm{e}\text{-}17$

(7) $\int_0^1 \int_0^1 16x^3 y^3 \, dx dy = 1$
Estimated error $= 1$
True error $= 7.9797\text{e-}18$

(8) Fill in the following table for the integral of the function $25x^4 y^4$ over the square $[0,1] \times [0,1]$.

n	integral	estimated error	true error
2	0.99978	2.1681e-04	2.1700e-04
4	0.99999	1.3563e-05	1.3563e-05
8	1.00000	8.4771e-07	8.4771e-07
16	1.00000	5.2982e-08	5.2982e-08

(9) Are your results consistent with the global order of accuracy of $O(h^4)$? (yes/no)

Solution 10.6.

(2) Integrate **three_peaks** over the square $[-1,1] \times [-1,1]$ and fill in the following table.

n	integral	estimated error	true error
10	1.7553	1.4612e-04	5.4743e-05
20	1.7552	4.6100e-06	7.2122e-08
40	1.7552	4.2841e-07	1.3603e-09
80	1.7552	2.6741e-08	8.5045e-11
160	1.7552	1.6985e-09	5.3211e-12

(3) Are the true error values consistent with the convergence rate of $O(h^4)$? (yes/no)

(4) Include a copy of your q_total_noabs.m here:

```
function [q,errest]=q_total_noabs(f,x,y,H,n)
  % [q,errest]=q_total(f,x,y,H,n)
  % n=number of intervals along one side
  % f=function handle for function to be integrated
  % x=lower left point abcissa
  % y=lower left point ordinate
  % H=length of side
  % q=integral of f
  % errest= error estimate

  % M. Sussman

  h=(H)/n;
  eltCount=0;
```

```
     for k=1:n
       for j=1:n
         eltCount=eltCount+1;
         elt(eltCount).x= x+(k-1)*h;
         elt(eltCount).y= y+(j-1)*h;
         elt(eltCount).h= h;
         [elt(eltCount).q,elt(eltCount).errest] = ...
             qerr_elt(f,elt(eltCount).x,elt(eltCount).y,h);
       end
     end
     if numel(elt) ~= n^2
       error('q_total: wrong number of elements!')
     end

     q=0;
     errest=0;
     for k=1:numel(elt);
       q=q+elt(k).q;
       errest=errest+   (elt(k).errest);
     end
   end
```

For $n = 80$, the integral of **three_peaks** over $[-1, 1] \times [-1, 1]$ is 1.7552
Estimated error $= 8.5015\mathrm{e}\text{-}11$
True error $= 8.5045\mathrm{e}\text{-}11$

Solution 10.7.

(2) Include your q_adaptive.m here:

```
function [q,errest,elt]=q_adaptive(f,x,y,H,tolerance)
  % [q,errest,elt]=q_adaptive(f,x,y,H,tolerance)
  % Uses an adaptive strategy over a square [x,x+H] X [y,y+H]
  % f is function handle for the integrand
  % x is lower-left abscissa
  % y is lewer-left ordinate
  % H is length of side
  % tolerance is desired accuracy

  % M. Sussman

  % your name and the date
  MAX_PASSES=500;
```

```
% initialize elt
elt(1).x=x;
elt(1).y=y;
elt(1).h=H;
[elt(1).q, elt(1).errest]=qerr_elt(f,x,y,H);

for passes=1:MAX_PASSES
  % compute q by adding up elemental values
  % and compute errest by adding up absolute elemental values
  % use a loop for this because the "sum" function cannot
  % be used for structures.
  q=0;
  errest=0;
  for k=1:numel(elt)
    q=q+elt(k).q;
    errest=errest+abs(elt(k).errest);
  end

  % if error meets tolerance, return
  if errest <= tolerance
    return;
  end

  % use a loop to find the element with largest abs(errest)
  maxerr=0;
  for k=1:numel(elt)
    if abs(elt(k).errest) > maxerr
      kmax=k;
      maxerr=abs(elt(k).errest);
    end
  end

  % replace that element with a quarter-sized element
  k=kmax;
  x=elt(k).x;
  y=elt(k).y;
  h=elt(k).h;

  % new values for this element
  elt(k).x=x+h/2;
  elt(k).y=y+h/2;
  elt(k).h=h/2;
```

```
[elt(k).q, elt(k).errest]= ...
    qerr_elt(f,elt(k).x,elt(k).y,elt(k).h);

% add three more quarter-sized elements
k=numel(elt)+1;
elt(k).x= x+h/2;
elt(k).y= y;
elt(k).h=h/2;
[elt(k).q, elt(k).errest]= ...
    qerr_elt(f,elt(k).x,elt(k).y,elt(k).h);

k=numel(elt)+1;
elt(k).x= x;
elt(k).y= y;
elt(k).h=h/2;
[elt(k).q, elt(k).errest]= ...
    qerr_elt(f,elt(k).x,elt(k).y,elt(k).h);

k=numel(elt)+1;
elt(k).x= x;
elt(k).y= y+h/2;
elt(k).h=h/2;
[elt(k).q, elt(k).errest]=qerr_elt(f,elt(k).x,elt(k).y,elt(k).h);

  end
  error('q_adaptive convergence failure.');
end
```

(3) Compute $\int_0^1 \int_0^1 16x^3 y^3 \, dx dy$
    integral $= 1$
    estimated error $= 0$
    exact error $= 0$

(4) Compute $\int_{-1}^1 \int_{-1}^1 16x^3 y^3 \, dx dy$
    integral $= 4$
    estimated error $= 5.9212e\text{-}17$
    exact error $= 0$

(5) Compute $\int_0^1 \int_0^1 25x^4 y^4 \, dx dy$
    integral $= 0.99978$
    estimated error $=2.1681e\text{-}04$
    exact error $= 2.1700e\text{-}04$
    numel(elt)=4? (yes/no)
    Include your plot here:

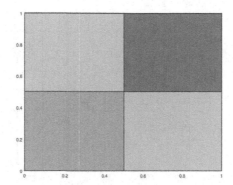

(6) Compute $\int_0^1 \int_0^1 25x^4y^4 \, dxdy$, finer tolerance
integral $= 0.99988$
estimated error $= 1.1832\text{e-}04$
exact error $= 1.1846\text{e-}04$
`numel(elt)=7?` (yes/no) Include your plot here:

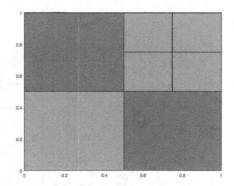

Solution 10.8.

(1) Compute $\int_0^1 \int_0^1 25x^4y^4 \, dxdy$, `tol=1.e-6`
integral $= 1.00000$
estimated error $= 9.5001\text{e-}07$
exact error $= 9.5015\text{e-}07$
`numel(elt)=?` 55
Include your plot here:

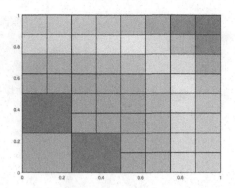

(2) Compute $\int_0^1 \int_0^1 25x^4y^4 \, dxdy$, `tol=9.e-7`
integral $= 1.00000$
estimated error $= 8.5076\text{e-}07$
exact error $= 8.5081\text{e-}07$
`numel(elt)=?` 61
Include your plot here:

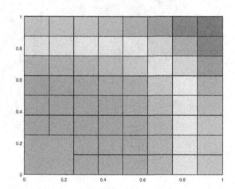

(3) Compute $\int_0^1 \int_0^1 25x^4y^4 \, dxdy$, `tol=5.e-7`
integral $= 1.00000$
estimated error $= 4.8220\text{e-}07$
exact error $= 4.8225\text{e-}07$
`numel(elt)=?` 97
Include your plot here:

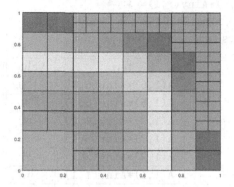

Solution 10.9.

Compute $\int_{-1}^{1} \int_{-1}^{1} \texttt{three\_peaks}(x, y)\, dx dy$, `tol=1.e-5`

integral = 1.7552

estimated error = 4.9911e-07

exact error = 2.6449e-07

`numel(elt)=?` 625

Include your plot here:

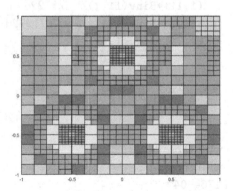

Solution 10.10.

(1) What is the result of `A*Ainv`? 6×6 identity matrix.

(2) What is the upper left 5×5 square of `C=B*Binv`?

4194304	4194304	131072	0	0
4194304	4194304	131072	0	0
4194304	4194304	131072	0	0
4194304	4194304	131072	0	0
4194304	4194304	131072	0	0

(3) Compute `A(1,:)*Ainv(:,1)= 1`
     `B(1,:)*Binv(:,1)=4194304`

(4) To see what goes right, compute the terms:

Term	Value
`A(1,6)*Ainv(6,1)`	-120
`A(1,5)*Ainv(5,1)`	240
`A(1,4)*Ainv(4,1)`	-180
`A(1,3)*Ainv(3,1)`	80
`A(1,2)*Ainv(2,1)`	-25
`A(1,1)*Ainv(1,1)`	6
Sum	1

(5) To see what goes wrong, compute the terms:

Term	Value
`B(1,24)*Binv(24,1)`	-2.5852e+22
`B(1,23)*Binv(23,1)`	5.1704e+22
`B(1,22)*Binv(22,1)`	-3.8778e+22
`B(1,21)*Binv(21,1)`	1.7235e+22
`B(1,20)*Binv(20,1)`	-5.3858e+21
`B(1,16)*Binv(16,1)`	-5.7705e+18
`B(1,11)*Binv(11,1)`	5.8122e+13
`B(1,6)*Binv(6,1)`	-76719720
`B(1,1)*Binv(1,1)`	24

Solution 10.11.

(2) `S=9997.8877`

(3) `a=9994.9707`

(4) `b=9997.8994`

(5) 3 digits of a agree with those of S
     7 digits of b agree with those of S

(6) `abs((a-S)/S)=2.9176e-04`
     `abs((b-S)/S)=1.1721e-06`

(7) $a_{1000} = 952.6$
     the value of the next term in the series, $x^{1001} = 0.90$

(8) For b, explain why roundoff error is so much smaller in this case.
     Adding smallest numbers first causes the sum to accumulate more slowly so that more terms are accurately treated before the sum becomes much larger than the next untreated term.

Differential Equations and Linear Algebra

Explicit ODE methods

Solution 11.1.

(3) Do your results agree with the first line of the table? (yes/no)
Fill in the table

Euler's explicit method				
numSteps	Step size (h)	Euler	Error	Ratio
10	0.2	-3.21474836	5.5922e-02	2.0323
20	0.1	-3.2432	2.7517e-02	2.0165
40	0.05	-3.2570	1.3646e-02	2.0083
80	0.025	-3.2639	6.7950e-03	2.0042
160	0.0125	-3.2673	3.3904e-03	2.0021
320	0.00625	-3.2690	1.6935e-03	—

(4) Estimated $p = 1$ (first order)

Solution 11.2.

(1) Include your rk2.m here:

```
function [ t, u ] = rk2 ( f_ode, tRange, uInitial, numSteps )
% [ t, u ] = rk2 ( f_ode, tRange, uInitial, numSteps )
% du/dt=f_ode(t,u).
% Runge-Kutta second order metthod
% f = function handle for a function with signature
%    fValue = f_ode(t,u)
% where fValue is a column vector
% tRange = [t1,t2] where the solution is sought on t1<=t<=t2
% uInitial = column vector of initial values for u at t1
% numSteps = number of equally-sized steps to take from t1 to t2
% t = row vector of values of t
% u = matrix whose k-th column is the approximate solution at t(k).
```

% M. Sussman

```
if size(uInitial,2) > 1
   error('rk2: uInitial must be scalar or a column vector.')
end

t(1,1) = tRange(1);
h = ( tRange(2) - tRange(1) ) / numSteps;
u(:,1) = uInitial;
for k = 1 : numSteps
   ta = t(k)+h/2;
   ua = u(:,k)+0.5*h*f_ode(t(k),u(:,k)) ;
   t(1,k+1) = t(1,k) + h;
   u(:,k+1) = u(:,k) + h * f_ode(ta,ua);
end
end
```

(2) Do your results agree with the first line of the table? (<u>yes</u>/no)
Fill in the table

numSteps	Step size (h)	RK2	RK2 Error	Ratio
10	0.2	-3.274896063	4.2255e-03	4.3367
20	0.1	-3.2716	9.7435e-04	4.1588
40	0.05	-3.2709	2.3429e-04	4.0771
80	0.025	-3.2707	5.7464e-05	4.0380
160	0.0125	-3.2707	1.4231e-05	4.0189
320	0.00625	-3.2707	3.5409e-06	—

(3) Estimated $p = 2$ (second order)
(4) Fill in the following table

numSteps	Step size (h)	Euler Error	RK2 Error
10	0.2	5.5922e-02	4.2255e-03
20	0.1	2.7517e-02	9.7435e-04
40	0.05	1.3646e-02	2.3429e-04
80	0.025	6.7950e-03	5.7464e-05
160	0.0125	3.3904e-03	1.4231e-05
320	0.00625	1.6935e-03	3.5409e-06

(5) Roughly how many steps? 160 or a little less
(6) Roughly how many steps would Euler require to achieve the accuracy that RK2 has for numSteps=320?
Explain your reasoning.

(i) Euler $p_E = 1$, so error=$C_E h$. For $h = 0.00625$, Euler error is 1.6935e-3, so $C_E = .0016935/0.00625 = 0.271$.

(ii) To reach a target error of 3.5409e-06, Euler would need $h = (3.5409\text{e-}06)/C_E = (3.5409\text{e-}06/0.271) = 1.3066\text{e-}05$

(iii) Number of steps would have be 2/1.3066e-05=1.5307e+05.

(7) What is the accuracy that Euler actually gets for your estimated number of steps? 3.5366e-06.

Is it about what you expected? (yes/no)

Solution 11.3.

(1) Include a copy of your rk3.m here

```
function [ t, u ] = rk3 ( f_ode, tRange, uInitial, numSteps )
  % [ t, u ] = rk3 ( f_ode, tRange, uInitial, numSteps )
  % du/dt=f_ode(t,u).
  % Runge-Kutta third order metthod
  % f = function handle for a function with signature
  %     fValue = f_ode(t,u)
  % where fValue is a column vector
  % tRange = [t1,t2] where the solution is sought on t1<=t<=t2
  % uInitial = column vector of initial values for u at t1
  % numSteps = number of equally-sized steps to take from t1 to t2
  % t = row vector of values of t
  % u = matrix whose k-th column is the approximate solution at t(k).

  % M. Sussman

  if size(uInitial,2) > 1
     error('rk2: uInitial must be scalar or a column vector.')
  end

  t(1,1) = tRange(1);
  h = ( tRange(2) - tRange(1) ) / numSteps;
  u(:,1) = uInitial;
  for k = 1 : numSteps
    ta = t(k)+h/2;
    ua = u(:,k)+0.5*h*f_ode(t(k),u(:,k)) ;
    tb = t(k)+h;
    ub = u(:,k)+h*(2*f_ode(ta,ua)-f_ode(t(k),u(:,k)));
    t(1,k+1) = t(1,k) + h;
    u(:,k+1) = u(:,k) + h * (f_ode(t(k),u(:,k))+4*f_ode(ta,ua)
    +f_ode(tb,ub))/6;
  end
end
```

(2) Do your results agree with the first line of the table? (yes/no)
Fill in the table

numSteps	Step size (h)	RK3	RK3 Error	Ratio
10	0.2	-3.27045877	2.1179e-04	8.6670
20	0.1	-3.2706	2.4437e-05	8.3270
40	0.05	-3.2707	2.9346e-06	8.1618
80	0.025	-3.2707	3.5956e-07	8.0804
160	0.0125	-3.2707	4.4497e-08	8.0401
320	0.00625	-3.2707	5.5344e-09	—

(3) Estimated $p = 8$
(4) Using some results from Exercise 11.2, fill in the following table:

numSteps	Step size (h)	RK2 Error	RK3 Error
10	0.2	4.2255e-03	2.1179e-04
20	0.1	9.7435e-04	2.4437e-05
40	0.05	2.3429e-04	2.9346e-06
80	0.025	5.7464e-05	3.5956e-07
160	0.0125	1.4231e-05	4.4497e-08
320	0.00625	3.5409e-06	5.5344e-09

(5) Roughly how many steps? 40
(6) Roughly how many steps would RK2 require to achieve the accuracy that RK3 has for numSteps=320?
Explain your reasoning.

    (i) $p_{RK2} = 2$, so error=$C_{RK2}h^2$. For $h = 0.00625$, RK2 error is 3.5409e-06, so $C_{RK2} = 3.5409\text{e-}06/(0.00625)^2 = 0.0906$.

    (ii) To reach a target error of 5.5344e-09, RK2 would need $h = \sqrt{5.5344\text{e-}09/.0906} = 2.4716\text{e-}04$.

    (iii) Number of steps would have be 2/2.4716e-04=8092.

(7) What is the accuracy that RK2 actually gets for your estimated number of steps? 5.5129e-09
Is it about what you expected? (yes/no)

Solution 11.4.

(1) What is fValue?

   -8
   -9

(2) What is usystem(:,end)? [-2.7293 ; -2.5940]
(3) u1(end)=-2.7293 and u2(end)=-2.5940

(4) Are both norms roundoff or zero? (yes/no)

Solution 11.5.

(1) Include a copy of **pendulum_ode.m** here:
(2) Fill in the first columns of the following table.
(3) Complete the following table:

t	n	Euler		RK3	
0.00	1	1.00	0.00	1.00	0.00
6.25	251	-1.06185	0.97913	-0.75431	1.06326
12.50	501	1.21137	-1.33835	0.11877	-1.64791
18.75	751	-1.82570	0.35516	0.58196	1.33043
25.00	1001	0.90952	2.72745	-0.97403	-0.35971

(4) There is no hint, beyond lack of conservation of energy, indicating that the Euler's method solution is not correct. Plots:

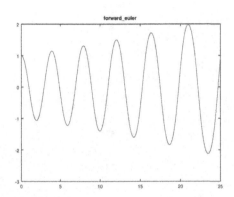 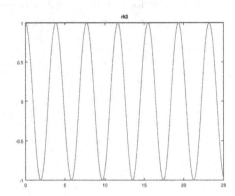

(5) Plot of RK3 and refined Euler solution:

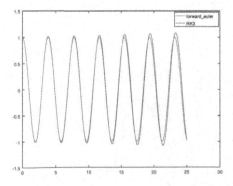

Solution 11.6.

(1) Step size is 0.5. Include your plot here:

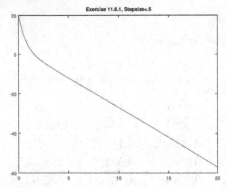

(2) Step size for `numSteps=30` is 0.6667
    Step size for `numSteps=20` is 1
    Step size for `numSteps=15` is 1.3333
    Step size for `numSteps=12` is 1.6667
    Step size for `numSteps=10` is 2

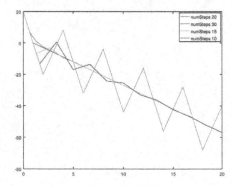

(3) Step size for `numSteps=8` is 2.5

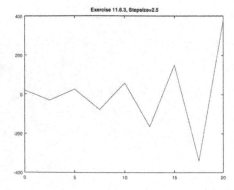

(4) Step size for `numSteps=20` is 1
   Step size for `numSteps=10` is 2
   Step size for `numSteps=9` is 2.2222
   Step size for `numSteps=8` is 2.5

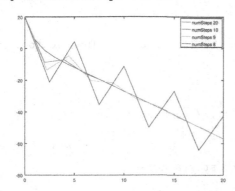

(5) Step size for `numSteps=7` is 2.8571

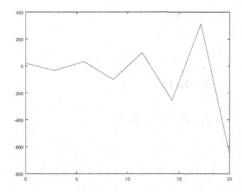

Solution 11.7.

(2) Include a copy of your modified `ab2.m` here:

```
function [ t, u ] = ab2 ( f_ode, tRange, uInitial, numSteps )
  % [ t, u ] = ab2 ( f_ode, tRange, uInitial, numSteps ) uses
  % Adams-Bashforth second-order method to solve a system
  % of first-order ODEs du/dt=f_ode(t,u).
  % f = name of an m-file with signature
  %     fValue = f_ode(t,u)
  % to compute the right side of the ODE as a column vector
  %
  % tRange = [t1,t2] where the solution is sought on t1<=t<=t2
  % uInitial = column vector of initial values for u at t1
  % numSteps = number of equally-sized steps to take from t1 to t2
```

```
% t = row vector of values of t
% u = matrix whose k-th row is the approximate solution at t(k).

% M. Sussman

if size(uInitial,2) > 1
   error('ab2: uInitial must be scalar or a column vector.')
end

t(1) = tRange(1);
h = ( tRange(2) - tRange(1) ) / numSteps;
u(:,1) = uInitial;

% The Runge-Kutta algorithm starts here
k = 1;
   fValue =  f_ode( t(k), u(:,k) );
   thalf = t(k) + 0.5 * h;
   uhalf = u(:,k) + 0.5 * h * fValue;
   fValuehalf = f_ode( thalf, uhalf );

   t(1,k+1) = t(1,k) + h;
   u(:,k+1) = u(:,k) + h * fValuehalf;

% The Adams-Bashforth algorithm starts here
for k = 2 : numSteps
   fValueold=fValue;
   fValue = f_ode( t(k), u(:,k) );
   t(1,k+1) = t(1,k) + h;
   u(:,k+1) = u(:,k) + h * ( 3 * fValue - fValueold ) / 2;
end
end
```

(3) Explain the role of fValue and fValueold
    fValueold is fvalue from the previous step. It is needed for the AB2 compu-
    tation of u at the current step.
(4) How many times is f_ode called? 101
(5) Fill in the following table:

numSteps	Step size (h)	AB2 ($t = 2$)	AB2 Error	Ratio
10	0.2	-3.28013993	9.4694e-03	4.0854
20	0.1	-3.2730	2.3179e-03	4.0519
40	0.05	-3.2712	5.7204e-04	4.0281
80	0.025	-3.2708	1.4201e-04	4.0145
160	0.0125	-3.2707	3.5375e-05	4.0074
320	0.00625	-3.2707	8.8273e-06	—

(6) Estimate $p = 2$

Solution 11.8.

(1) Include a copy of your heun.m here:

```
function [ t, u ] = heun ( f_ode, tRange, uInitial, numSteps )
  % [ t, u ] = heun ( f_ode, tRange, uInitial, numSteps )
  % du/dt=f_ode(t,u).
  % Heun metthod
  % f = function handle for a function with signature
  %     fValue = f_ode(t,u)
  % where fValue is a column vector
  % tRange = [t1,t2] where the solution is sought on t1<=t<=t2
  % uInitial = column vector of initial values for u at t1
  % numSteps = number of equally-sized steps to take from t1 to t2
  % t = row vector of values of t
  % u = matrix whose k-th column is the approximate solution at t(k).

  % M. Sussman

  if size(uInitial,2) > 1
    error('heun: uInitial must be scalar or a column vector.')
  end


  t(1,1) = tRange(1);
  h = ( tRange(2) - tRange(1) ) / numSteps;
  u(:,1) = uInitial;
  for k = 1 : numSteps
    fk = f_ode(t(k),u(:,k));
    ua = u(:,k)+h*fk;
    t(1,k+1) = t(1,k) + h;
    u(:,k+1) = u(:,k) + h/2 * (fk + f_ode(t(k+1),ua) );
  end
end
```

(2) Fill in the first columns of the following table. Ratios of errors = 4.2964, 4.1423, 4.0694, 4.0343, 4.0170, so p=2.

(3) Complete the following table:

	Heun		AB2		RK2	
numSteps	Error	Calls	Error	Calls	Error	Calls
10	1.2784e-02	20	8.5772e-03	11	4.9118e-03	20
20	2.9755e-03	40	1.7899e-03	21	1.1768e-03	40
40	7.1833e-04	80	3.9601e-04	41	2.8679e-04	80
80	1.7652e-04	160	9.2172e-05	81	7.0731e-05	160
160	4.3754e-05	320	2.2166e-05	161	1.7561e-05	320
320	1.0892e-05	640	5.4307e-06	321	4.3748e-06	640

(4) What is the order of accuracy of Heun's method? 2
What is the order of accuracy of AB2? 2
What is the order of accuracy of RK2? 2

(5) Comparing Heun's method with AB2, which appears to have the greater accuracy for a given step size? AB2
Assuming that each function call takes a relatively long time, which of these two methods would be expected to yield the greater accuracy in the lesser time? AB2

(6) When RK2 is included, which of the three would be the best choice based on accuracy for a given step size? RK2
Based on greatest accuracy in the least time? AB2

Solution 11.9.

(1) Include your script here:

```
% Exercise 11.9, plotting stability regions in the complex plane
% stab1.m
% M. Sussman

% part (a)
theta=linspace(0,2*pi,1000);
zeta=exp(i*theta);

% check: on unit circle?
maxradius=max(abs(zeta));
minradius=min(abs(zeta));
if norm( [maxradius-1,minradius-1] )>1.e-12
```

```
   error('stab.m: this cannot happen!')
end

% part (b): explicit Euler
mu=zeta-1;

%plot(zeta) % for part (a)
plot(mu,'b') % for part (b)
axis('equal')
hold on

% horizontal axis (make sure it goes into right half plane
plot([min(real(mu)),max(max(real(mu)),.05)]*(1.1+eps*i),'k')
% vertical axis
plot([min(imag(mu)),max(imag(mu))]*(1.1*i),'k')

% add the inner curve: part (c)
mu=0.95*zeta-1;
plot(mu,'c')

hold off
```

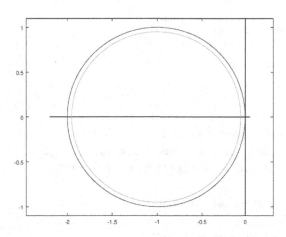

(2) Include your script to display both stability regions here:

```
% Exercise 11.9, plotting stability regions in the complex plane
% stab2.m
% M. Sussman

% part (a)
```

```
theta=linspace(0,2*pi,1000);
zeta=exp(i*theta);

% check: on unit circle?
maxradius=max(abs(zeta));
minradius=min(abs(zeta));
if norm( [maxradius-1,minradius-1] )>1.e-12
  error('stab.m: this cannot happen!')
end

% part (b): explicit Euler
mu=zeta-1;

%plot(zeta) % for part (a)
plot(mu,'b') % for part (b)
axis('equal')
hold on

% horizontal axis (make sure it goes into right half plane
plot([min(real(mu)),max(max(real(mu)),.05)]*(1.1+eps*i),'k')
% vertical axis
plot([min(imag(mu)),max(imag(mu))]*(1.1*i),'k')

% add the inner curve: part (c)
mu95=0.95*zeta-1;
zeta95=zeta*.95;
mu95=zeta95-1;
plot(mu95,'c')


% now do the stability region for ab2
% u(k+1) - (1+(3*h*lambda*u(k)-h*lambda*u(k-1))/2) = 0
% divide by u(k-1) and set zeta=u(k+1)/u(k) for all k
% set mu=h*lambda
% 2*zeta^2 - 2*zeta - 3*mu*zeta + mu=0
% solve for mu
% 2*zeta^2 - 2*zeta + mu*(-3*zeta+1)=0
% mu = 2*(zeta^2-1)/(1-3*zeta)
mu = 2*(zeta-zeta.^2)./(1-3*zeta);
plot(mu,'r')
mu95 = 2*(zeta95-zeta95.^2)./(1-3*zeta95);
plot(mu95,'y')
```

```
hold off
```

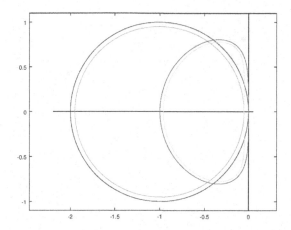

(3) Give examples, with explanation, of odes and values of h that:

(a) Would be stable using both **ab2** and explicit Euler

$h\lambda = -0.5$ is inside both regions, so $h = 1$ for the ODE $du/dt = -u/2$.

(b) Would be stable using **ab2** but not for explicit Euler

$h\lambda = -.05 \pm .5i$ is outside the Euler stability region but inside the AB2 region, so $du/dt = -(1 + 10i)$ with $h = 1/20$ does it. In addition, the system

$$\frac{du_1}{dt} = u_2$$

$$\frac{du_2}{dt} = -101u_1 - 2u_2$$

with $h = 1/20$ is stable for AB2 but not for explicit Euler.

(c) Would be stable using explicit Euler but not **ab2**

$h\lambda = -1.5$ is inside the Euler stability region and outside the AB2 region, so $du/dt = -1.5u$ with $h = 1$ does it.

Implicit ODE methods

Solution 12.1.

(3) Include a copy of your stiff10000_ode.m here:

```
function fValue = stiff10000_ode ( t, u )
  % fValue = stiff10000_ode ( t, u )
  % computes the right side of the ODE
  %   du/dt=f_ode(t,u)=lambda*(-u+sin(t)) for lambda = 10000
  % t is independent variable
  % u is dependent variable
  % output, fValue is the value of f_ode(t,u).

  % M. Sussman

  LAMBDA=10000;
  fValue = LAMBDA * ( -u + sin(t) );
end
```

Include a copy of your stiff10000_solution.m here:

```
function u = stiff10000_solution ( t )
  % u = stiff10000_solution ( t )
  % computes the solution of the ODE
  %   du/dt=f_ode(t,u)=lambda*(-u+sin(t)) for lambda = 10000
  % and initial condition u=0 at t=0
  % t is the independent variable
  % u is the solution value

  % M. Sussman

  LAMBDA=10000;
  u = (LAMBDA^2/(1+LAMBDA^2))*sin(t) + ...
```

```
            (LAMBDA  /(1+LAMBDA^2))*(exp(-LAMBDA*t)-cos(t));
  end
```

(4) Nothing required.
(5) Include a copy of your direction field plot plus solution here:

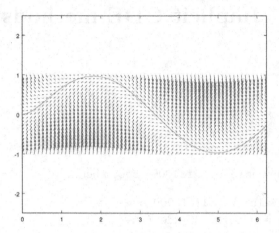

Solution 12.2.

(1) Do you agree that 40 points is about right? (<u>yes</u>/no)
(2) `forward_euler.m` required 40000 steps
    `rk3.m` also required 40000 steps

Solution 12.3.

(1) Include a copy of your `back_euler_lam.m` here:

```
function [t,u]=back_euler_lam(lambda,tRange,uInitial,numSteps)
  % [t,u]=back_euler_lam(lambda,tRange,uInitial,numSteps)
  % Euler's implicit method to solve a particular ODE
  % du/dt=lambda*u + sin(t)
  % tRange = [t1,t2] where the solution is sought on t1<=t<=t2
  % uInitial = column vector of initial values for u at t1
  % numSteps = number of equally-sized steps to take from t1 to t2
  % t = row vector of values of t
  % u = matrix whose k-th column is the approximate solution at t(k).

  % M. Sussman

  if size(uInitial,2) > 1
    error('back_euler_lam: uInitial must be scalar or
    a column vector.')
  end
```

```
    t(1) = tRange(1);
    h = ( tRange(2) - tRange(1) ) / numSteps;
    u(:,1) = uInitial;
    for k = 1 : numSteps
      t(1,k+1) = t(1,k) + h;
      u(:,k+1) = (u(:,k) + h * lambda * sin(t(k+1)))/(1+h*lambda);
    end
  end
```

(2) Does your solution blow up? (yes/<u>no</u>)
(3) Include your plot here:

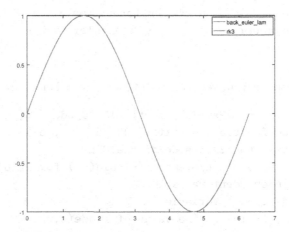

Are the two solutions close? (<u>yes</u>/no) (lines on top of each other)
(4) Did you get the provided values? (<u>yes</u>/no)

Solution 12.4.

(1) Fill in the table:

lambda=55		
numSteps	back_euler_lam error	ratio
40	1.2607e-04	2.8300
80	4.4547e-05	2.5269
160	1.7629e-05	2.3041
320	7.6513e-06	2.1647
640	3.5346e-06	2.0859
1280	1.6945e-06	2.0439
2560	8.2903e-07	—

(2) The order of accuracy is 1.
(3) Fill in the following table;

forward_euler error comparison		
numSteps	forward_euler error	ratio
40	unstable	
80	unstable	
160	unstable	
320	5.3206e-06	1.8025
640	2.9519e-06	1.9059
1280	1.5488e-06	1.9540
2560	7.9262e-07	—

(4) What is the order of accuracy using forward_euler? 1
What is the order of accuracy using back_euler_lam? 1

Solution 12.5.

(1) Include a copy of your revised stiff10000_ode.m here:

```
function [fValue,fPartial] = stiff10000_ode ( t, u )
  % [fValue,fPartial] = stiff10000_ode ( t, u )
  % computes the right side of the ODE
  %   du/dt=f_ode(t,u)=lambda*(-u+sin(t)) for lambda = 10000
  % t is independent variable
  % u is dependent variable
  % output, fValue is the value of f_ode(t,u),
  % fPartial is the value of the derivative d(fValue)/du

  % M. Sussman

  LAMBDA=10000;
  fValue = LAMBDA * ( -u + sin(t) );
  fPartial = -LAMBDA;
end
```

(2) Include a copy of your revised stiff55_ode.m here:

```
function [fValue,fPartial] = stiff55_ode ( t, u )
  % [fValue,fPartial] = stiff55_ode ( t, u )
  % computes the right side of the ODE
  %   du/dt=f_ode(t,u)=lambda*(-u+sin(t)) for lambda = 55
  % t is independent variable
  % u is dependent variable
  % output, fValue is the value of f_ode(t,u).
```

```
   % fPartial is the value of the derivative d(fValue)/du

   % M. Sussman

   LAMBDA=55;
   fValue = LAMBDA * ( -u + sin(t) );
   fPartial = -LAMBDA;
end
```

(5) Include a copy of your back_euler.m, including comments, here:

```
function [t,u]=back_euler(f_ode,tRange,uInitial,numSteps)
   % [t,u]=back_euler(f_ode,tRange,uInitial,numSteps) computes
   % the solution to an ODE by the backward Euler method
   % f_ode is the handle of a function whose signature is
   %      [fValue,fPartial]=f_ode(t,u)
   % tRange is a two dimensional vector of beginning and
   %    final values for t
   % uInitial is a column vector for the initial value of u
   % numSteps is the number of evenly-spaced steps to divide
   %    up the interval tRange
   % t is a row vector of selected values for the
   %    independent variable
   % u is a matrix. The k-th column of u is
   %    the approximate solution at t(k)

   % M. Sussman

   % force t to be a row vector
   t(1,1) = tRange(1);
   h = ( tRange(2) - tRange(1) ) / numSteps;
   u(:,1) = uInitial;
   % the following for loop performs the spatial stepping (on t)
   for k = 1 : numSteps
     t(1,k+1) = t(1,k) + h;

     % The following statement computes the initial guess for Newton
     U = (u(:,k)) + h * f_ode( t(1,k), u(:,k));
     [U,isConverged]= newton4euler(f_ode,t(k+1),u(:,k),U,h);

     if ~ isConverged
       error(['back_euler failed to converge at step ', ...
             num2str(k)])
```

```
      end

      u(:,k+1) = U;
    end
end
```

Include a copy of your `newton4euler.m`, including comments, here:

```
function [U,isConverged]=newton4euler(f_ode,tkp1,uk,U,h)
  % [U,isConverged]=newton4euler(f_ode,tkp1,uk,U,h)
  % special function to evaluate Newton's method for back_euler
  % f_ode is the handle of a function whose signature is
  %    [fValue,fPartial]=f_ode(t,u)

  % M. Sussman

  TOL = 1.e-6;   % Convergence tolerance
  MAXITS = 500;  % maximum number of iterations before failure

  isConverged=false;
  for n=1:MAXITS
    [fValue fPartial] = f_ode( tkp1, U);
    F = uk + h * fValue - U;  % this statement implements (12.6)
      % The MATLAB function "eye" generates the identity matrix
    J = h * fPartial - eye(numel(U));
      % When F is a column vector and J a matrix,
      % the expression J\F means (J)^(-1)*F
    increment=J\F;  % This partially implements (12.5)
    if n>1
      r1=norm(increment,inf)/oldIncrement;
    else
      r1=2; % forces convergence failure
    end
    oldIncrement=norm(increment,inf);
    U = U - increment; % This partially implements (12.5)
    if norm(increment,inf) < TOL*norm(U,inf)*(1-r1)
      isConverged=true;  % turns TRUE when converged
      return
    end
  end
end
```

(6) What are the values of $\mathbf{J(U)} = \partial\mathbf{F}/\partial\mathbf{U}$ and $\mathbf{F(U)}$ as a MATLAB matrix and vector, respectively.

- $F_1 = 4U_1 + 2(U_2)^2 = 4*(-2) + 2*1^2 = -6$
  $F_2 = (U_1)^3 + 5U_2 = (-2)^3 + 5*1 = -3$
  So $\mathbf{F} = \begin{bmatrix} -6 \\ -3 \end{bmatrix}$.
- $\mathbf{J}_{ij} = \partial F_i/U_j$
  $F_{11} = 4$, $F_{21} = 3*U_1^2 = 12$, $F_{12} = 4U_2 = 4$, $F_{22} = 5$, so that

$$\mathbf{J} = \begin{bmatrix} 4 & 4 \\ 12 & 5 \end{bmatrix}$$

(7) What is the name of the variable that represents $\mathbf{J}^{(n)}$? J What is the name of the variable that represents $\Delta\mathbf{U}$? increment What is the name of the variable that represents $\mathbf{F}(\mathbf{U}^{(n)})$? fValue

(8) Is U a row vector or a column vector? (row/<u>column</u>)

(9) What is the value of $\mathbf{U}$? (at least 8 significant digits)

$$0 = u_k + h f_{\text{ODE}}(t_{k+1}, \mathbf{U}) - \mathbf{U}$$
$$0 = u_k + h\lambda(-\mathbf{U} + \sin(t_{k+1}) - \mathbf{U}$$
$$(\mathbf{U})(h\lambda + 1) = u_k + h\lambda\sin(t_{k+1})$$
$$\mathbf{U} = \frac{u_k + h\lambda\sin(t_{k+1})}{h\lambda + 1}$$
$$\mathbf{U} = \frac{1.0 + 0.1*10000*\sin(2.0)}{0.1*10000 + 1}$$
$$\mathbf{U} = 0.909388038786895$$

(10) What is the value `newton4euler` yields?

```
newton4euler(@stiff10000_ode,2.0,1.0,1,0.1)
ans =   0.909388038786895  %(same!)
```

Does it take only 2 iterations? (<u>yes</u>/no)

(11) What is the norm of the difference between the two solutions?

```
[t,ub]=back_euler(@stiff10000_ode,[0,2*pi],0,40);
[t,ulam]=back_euler_lam(10000,[0,2*pi],0,40);
norm(ub-ulam)
ans =   3.95456081509184e-16
```

Solution 12.6.

(1) Include your copy of `vanderpol_ode.m` here:

```
function [fValue, J]=vanderpol_ode(t,u)
  % [fValue, J]=vanderpol_ode(t,u)
  % more comments

  % M. Sussman
```

```
if size(u,1) ~=2  | size(u,2) ~=1
  error('vanderpol_ode: u must be a column vector of length 2!')
end

a=11;

fValue = [ u(2)
   -a*(u(1)^2-1)*u(2)-u(1)+exp(-t)];

df1du1 = 0;
df1du2 = 1;
df2du1 = -2*a*u(1)*u(2)-1;
df2du2 = -a*(u(1)^2-1);

J=[df1du1 df1du2
   df2du1 df2du2];

end
```

(2) Include your plot of the 40 interval here:

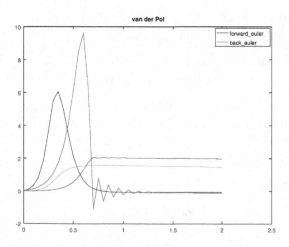

(3) Include your plot of the 640 interval here:

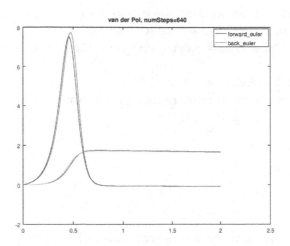

Solution 12.7.

(1) Include a copy of your `trapezoid.m` here:

```
function [t,u]=trapezoid(f_ode,tRange,uInitial,numSteps)
  % [t,u]=trapezoid(f_ode,tRange,uInitial,numSteps) computes
  % the solution to an ODE by the trapezoid or Crank-Nicolson method
  % f_ode is the handle of a function whose signature is
  %    [fValue,fPartial]=f_ode(t,u)
  % tRange is a two dimensional vector of beginning and
  %    final values for t
  % uInitial is a column vector for the initial value of u
  % numSteps is the number of evenly-spaced steps to divide
  %    up the interval tRange
  % t is a row vector of selected values for the
  %    independent variable
  % u is a matrix. The k-th column of u is
  %    the approximate solution at t(k)

  % M. Sussman

  % force t to be a row vector
  t(1,1) = tRange(1);
  h = ( tRange(2) - tRange(1) ) / numSteps;
  u(:,1) = uInitial;
  % the following for loop performs the spatial stepping (on t)
  for k = 1 : numSteps
    t(1,k+1) = t(1,k) + h;
```

```
% The following statement computes the initial guess for Newton
U = (u(:,k)) + h * f_ode( t(1,k), u(:,k));
[U,isConverged]= newton4trapezoid(f_ode,t(1,k),t(k+1),u(:,k),U,h);

    if ~ isConverged
      error(['trapezoid failed to converge at step ', ...
            num2str(k)])
    end

    u(:,k+1) = U;
  end
end
```

(2) Include a copy of your `newton4trapezoid.m` here:

```
function [U,isConverged]=newton4trapezoid(f_ode,tk,tkp1,uk,U,h)
  % [U,isConverged]=newton4trapezoid(f_ode,tk,tkp1,uk,U,h)
  % special function to evaluate Newton's method for trapezoid
  % f_ode is the handle of a function whose signature is
  %    [fValue,fPartial]=f_ode(t,u)

  % M. Sussman

  TOL = 1.e-6;  % Convergence tolerance
  MAXITS = 500; % maximum number of iterations before failure

  isConverged=false;
  fValuek = f_ode( tk, uk);
  for n=1:MAXITS
    [fValue fPartial] = f_ode( tkp1, U);
    F = uk + h/2 * (fValuek+fValue) - U;
      % The MATLAB function "eye" generates the identity matrix
    J = h/2 * fPartial - eye(numel(U));
      % When F is a column vector and J a matrix,
      % the expression J\F means (J)^(-1)*F
    increment=J\F;  % This partially implements (12.5)
    if n>1
      r1=norm(increment,inf)/oldIncrement;
    else
      r1=2; % forces convergence failure
    end
    oldIncrement=norm(increment,inf);
    U = U - increment; % This partially implements (12.5)
```

```
      if norm(increment,inf) < TOL*norm(U,inf)*(1-r1)
        isConverged=true;   % turns TRUE when converged
        return
      end
    end
  end
```

(3) What does `newton4trapezoid` give? 0.894104843976271
Does it take only 2 iterations? (yes/no)

(4) What does your hand calculation give?

$$0 = u_k + \frac{h}{2}(f_{\mathrm{ODE}}(t_k, u_k) + f_{\mathrm{ODE}}(t_{k+1}, \mathbf{U})) - \mathbf{U}$$

$$0 = u_k + \frac{h\lambda}{2}(-u_k + \sin(t_k) - \mathbf{U} + \sin(t_{k+1})) - \mathbf{U}$$

$$\mathbf{U} = \frac{u_k(1 - h\lambda/2) + h\lambda/2(\sin(t_k) + \sin(t_{k+1}))}{1 + h\lambda/2}$$

$$\mathbf{U} = \frac{1.0 + 0.1 * 55/2 * (\sin(1.9) + \sin(2.0))}{1 + 0.1 * 55/2}$$

$$\mathbf{U} = 0.894104843976271$$

Do they agree? (yes/no)

(5) Fill in the following table:

stiff55		
numSteps	trapezoid error	ratio
10	5.2779e-03	217.98972
20	2.4212e-05	0.64668
40	3.7440e-05	4.00740
80	9.3427e-06	4.00185
160	2.3346e-06	4.00046
320	5.8358e-07	—

(6) Are your results consistent with the theoretical $O(h^2)$ convergence rate of the trapezoid rule? (yes/no)

Solution 12.8.

(1) Include your plot here:

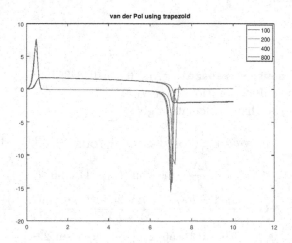

(2) Include your plot here:

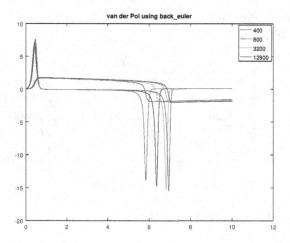

(3) Include your plot here:

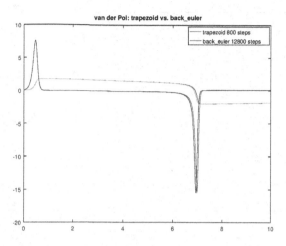

Solution 12.9.

(1) Include your plot here:

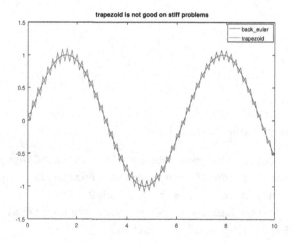

(2)

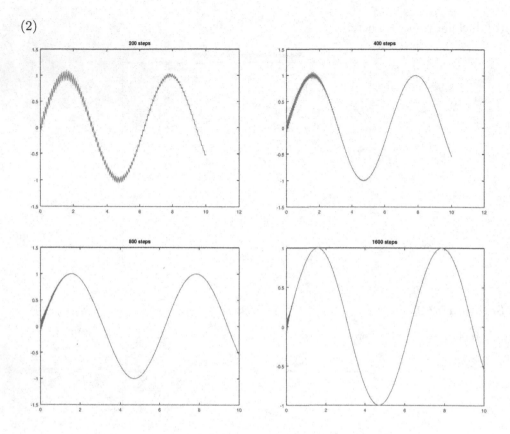

Solution 12.10.

(1) Include a copy of your **bdm2.m** here:

```
function [t,u]=bdm2(f_ode,tRange,uInitial,numSteps)
  % [t,u]=bdm2(f_ode,tRange,uInitial,numSteps) computes
  % the solution to an ODE by the bdm2 method
  % f_ode is the handle of a function whose signature is
  %    [fValue,fPartial]=f_ode(t,u)
  % tRange is a two dimensional vector of beginning and
  %    final values for t
  % uInitial is a column vector for the initial value of u
  % numSteps is the number of evenly-spaced steps to divide
  %    up the interval tRange
  % t is a row vector of selected values for the
  %    independent variable
  % u is a matrix. The k-th column of u is
  %    the approximate solution at t(k)
```

```
% M. Sussman

% force t to be a row vector
t(1,1) = tRange(1);
h = ( tRange(2) - tRange(1) ) / numSteps;
u(:,1) = uInitial;
% begin with a single backward Euler step
  k=1;
  t(1,k+1) = t(1,k) + h;

  % The following statement computes the initial guess for Newton
  U = (u(:,k)) + h * f_ode( t(1,k), u(:,k));
  [U,isConverged]= newton4euler(f_ode,t(k+1),u(:,k),U,h);

  if ~ isConverged
    error(['bdm2 failed to converge at step ', ...
        num2str(k)])
  end

  u(:,k+1) = U;
% next, bdm2 steps
for k = 2 : numSteps
  t(1,k+1) = t(1,k) + h;

  % The following statement computes the initial guess for Newton
  U = (u(:,k)) + h * f_ode( t(1,k), u(:,k));

  term=4*u(:,k)/3-u(:,k-1)/3;
  [U,isConverged]= newton4euler(f_ode,t(k+1),term,U,2*h/3);

  if ~ isConverged
    error(['bdm2: failed to converge at step ', ...
        num2str(k)])
  end

  u(:,k+1) = U;
  end
end
```

(2) Show that convergence is $O(h^2)$ and support your work.

```
sizes=[40 80 160 320];
for k=1:4;
```

```
    [t,u]=bdm2(@stiff55_ode,[0,2*pi],0,sizes(k));
    errors(k)=u(end)-stiff55_solution(2*pi);
end
errors =
   -1.4746e-04  -3.7187e-05  -9.3219e-06  -2.3327e-06
ratios=errors(1:end-1)./errors(2:end)
ratios =
     3.9654    3.9892    3.9962
```

Ratios are about 4, order of accuracy is 2.

(3) Include your plot here:

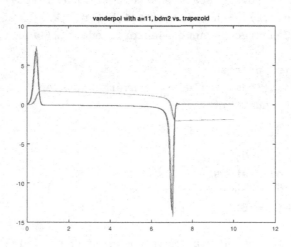

(4) Include your plot here:

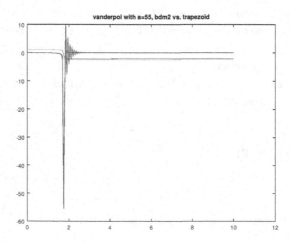

Solution 12.11.

(1) Include your plot for the final solution here:

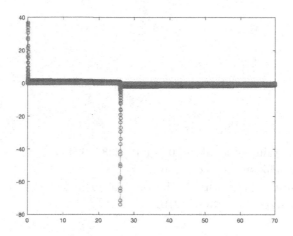

(2) Include your plot of the final solution from **ode45** here:

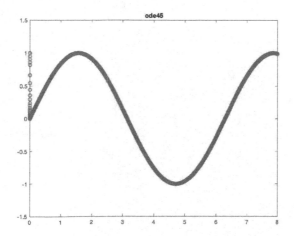

Include your plot of the final solution from `ode15s` here:

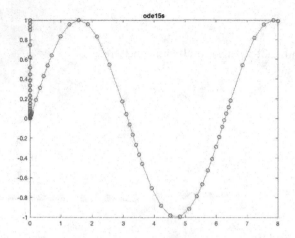

(3) What is the value of u_1 at x=70? -1.128238369949555

How many steps did it take? 256

(4) For tolerance of 1.e-8, what is the value of u_1 at x=70? -1.125147778949563

How many steps did it take? 2102

Chapter 13

Boundary value problems and partial differential equations

Solution 13.1.

(1) What is your solution U?

U =[1.00000; 0.36362; 0.12893; 0.26023; 0.72636; 1.50000]

(2) Include a copy of your `exer1b.m` here:

```
% Chapter 13, exercise 1
% File named exer1b.m
% M. Sussman

N = 4;
C = 0.05;
RHOG = 0.4;
% N interior mesh points, N+1 intervals
dx = 5.0 / ( N + 1 );
x = dx * (0:N+1);
A = [ -2*(1+C*dx)     +(1+1.5*C*dx)        0           0;
        +(1+1.5*C*dx) -2*(1+2*C*dx)   +(1+2.5*C*dx)     0;
              0        +(1+2.5*C*dx)  -2*(1+3*C*dx)  (1+3.5*C*dx);
              0             0          +(1+3.5*C*dx) -2*(1+4*C*dx) ];
ULeft=1;
URight=1.5;
b = [ RHOG*dx^2-(1+0.5*C*dx)*ULeft
      RHOG*dx^2
      RHOG*dx^2
      RHOG*dx^2-(1+4.5*C*dx)*URight];
U = A \ b;
U = [ULeft; U; URight]

% check second equation:
(1+1.5*C*dx)*U(2)-2*(1+2*C*dx)*U(3)+(1+2.5*C*dx)*U(4)-.4*dx^2
```

Is your result essentially zero? (<u>yes</u>/no)

(3) Include a copy of your `exer1c.m` here:

```
% Chapter 13, exercise 1
% File named exer1c.m
% M. Sussman


N = 4;
C = 0.05;
% RHOG = 0.4;
RHOG = 0;
% N interior mesh points, N+1 intervals
dx = 5.0 / ( N + 1 );
x = dx * (0:N+1);
A = [ -2*(1+C*dx)      +(1+1.5*C*dx)           0              0;
        +(1+1.5*C*dx)  -2*(1+2*C*dx)   +(1+2.5*C*dx)          0;
             0         +(1+2.5*C*dx)   -2*(1+3*C*dx)  (1+3.5*C*dx);
             0              0          +(1+3.5*C*dx) -2*(1+4*C*dx) ];
ULeft=1;
%URight=1.5;
URight=1.0;
b = [ RHOG*dx^2-(1+0.5*C*dx)*ULeft
      RHOG*dx^2
      RHOG*dx^2
      RHOG*dx^2-(1+4.5*C*dx)*URight];
U = A \ b;
U = [ULeft; U; URight]


norm(U-ones(6,1))
```

Is your result essentially correct? (<u>yes</u>/no)

(4) Include a copy of your `exer1d.m` here:

```
% Chapter 13, exercise 1
% File named exer1d.m
% M. Sussman


N = 4;
C = 0.05;
% RHOG = 0.4;
RHOG = C;
% N interior mesh points, N+1 intervals
dx = 5.0 / ( N + 1 );
x = dx * (0:N+1);
```

```
A = [ -2*(1+C*dx)      +(1+1.5*C*dx)          0              0;
         +(1+1.5*C*dx)  -2*(1+2*C*dx)   +(1+2.5*C*dx)         0;
              0          +(1+2.5*C*dx)  -2*(1+3*C*dx)  (1+3.5*C*dx);
              0               0          +(1+3.5*C*dx) -2*(1+4*C*dx) ];
%ULeft=1;
ULeft=0;
%URight=1.5;
URight=5.0;
b = [ RHOG*dx^2-(1+0.5*C*dx)*ULeft
      RHOG*dx^2
      RHOG*dx^2
      RHOG*dx^2-(1+4.5*C*dx)*URight];
U = A \ b;
U = [ULeft; U; URight]


norm(U-x')
```

Is your result essentially correct? (yes/no)

(5) Include a copy of your **exer1e.m** here:

```
% Chapter 13, exercise 1
% File named exer1e.m
% M. Sussman


N = 4;
C = 0.05;
% RHOG = 0.4;
% RHOG = C;
% N interior mesh points, N+1 intervals
dx = 5.0 / ( N + 1 );
x = dx * (0:N+1);
A = [ -2*(1+C*dx)      +(1+1.5*C*dx)          0              0;
         +(1+1.5*C*dx)  -2*(1+2*C*dx)   +(1+2.5*C*dx)         0;
              0          +(1+2.5*C*dx)  -2*(1+3*C*dx)  (1+3.5*C*dx);
              0               0          +(1+3.5*C*dx) -2*(1+4*C*dx) ];
%ULeft=1;
ULeft=0;
%URight=1.5;
URight=25.0;
RHOG = 2+4*C*x(2:5);
b = [ RHOG(1)*dx^2-(1+0.5*C*dx)*ULeft
      RHOG(2)*dx^2
      RHOG(3)*dx^2
```

```
          RHOG(4)*dx^2-(1+4.5*C*dx)*URight];
U = A \ b;
U = [ULeft; U; URight]
```

```
norm(U-(x')).^2)
```

Is your result essentially correct? (yes/no)

(6) Include a copy of your plot here:

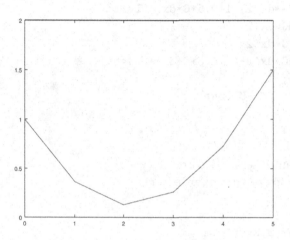

Is your result essentially correct? (yes/no)

Solution 13.2.

(1) Include a copy of your **rope_bvp**. here:

```
function [x,U]=rope_bvp(N)
  % [x,U]=rope_bvp(N)
  % Solves the system (13.3) but for arbitrary N
  % x is the space variable
  % U is the solution

  % M. Sussman

  C = 0.05;
  RHOG = 0.4;
  ULeft=1;
  URight=1.5;

  % N interior mesh points, N+1 intervals
  dx = 5.0 / ( N + 1 );
```

```
x = dx * (0:N+1);
A = zeros(N,N);
for k=1:N
  A(k,k) = -2*(1+C*k*dx);
end
for k=1:N-1
  A(k,k+1)=1+(k+.5)*C*dx;
  A(k+1,k)=1+(k+.5)*C*dx;
end
b = ones(N,1)*RHOG*dx^2;
b(1) = b(1) - (1+.5*C*dx)*ULeft;
b(N) = b(N) - (1+(N+.5)*C*dx)*URight;;
U = A \ b;
U = [ULeft; U; URight];
end
```

(2) Is U−U1 the zero vector? (<u>yes</u>/no)
(3) Does your modified copy yield correct results for $u = 1$? (<u>yes</u>/no)
(4) Does your modified copy yield correct results for $u = x$? (<u>yes</u>/no)
(5) Does your modified copy yield correct results for $u = x^2$? (<u>yes</u>/no)
(6) Include your plot here:

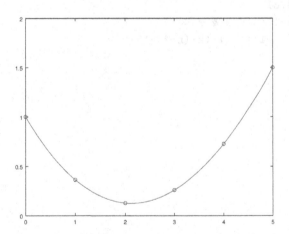

What is U(50)? U(50)=0.126957483290832

Solution 13.3.

(1) Include a copy of your phi.m here:

```
function z=phi(n,h,x)
% z=phi(n,h,x)
```

```
% Lagrange quadratic basis functions

% M. Sussman

if numel(x) > 1
  error('x is a scalar, not a vector, in phi.m');
end

if mod(n,2)==0   % n is even

  % (13.7)
  if (n-2)*h < x & x <= n*h
    z=(x-(n-1)*h)*(x-(n-2)*h)/(2*h^2);
  elseif n*h < x & x <= (n+2)*h
    z=((n+1)*h-x)*((n+2)*h-x)/(2*h^2);
  else
    z=0;
  end

else  % n is odd

  % (13.8)
  if (n-1)*h < x & x <= (n+1)*h
    z=((n+1)*h-x)*(x-(n-1)*h)/h^2;
  else
    z=0;
  end

end
end
```

(2) Include your plot here:

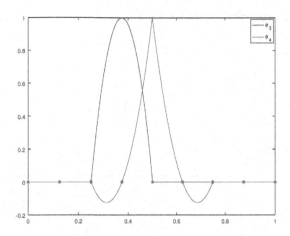

(3) Is $\phi_n(x)$ a continuous function? (yes/no)

(4) Include a copy of your `phip.m` here:

```
function z=phip(n,h,x)
% z=phip(n,h,x)
% derivative of Lagrange quadratic basis functions

% M. Sussman

if numel(x) > 1
   error('x is a scalar, not a vector, in phip.m');
end

if mod(n,2)==0   % n is even

   % (13.7)
   if (n-2)*h < x & x <= n*h
     z=((x-(n-1)*h)+(x-(n-2)*h))/(2*h^2);
   elseif n*h < x & x <= (n+2)*h
     z=(-((n+1)*h-x)-((n+2)*h-x))/(2*h^2);
   else
     z=0;
   end

else  % n is odd
```

```
% (13.8)
if (n-1)*h < x & x <= (n+1)*h
   z=(((n+1)*h-x)-(x-(n-1)*h))/h^2;
else
   z=0;
end


   end
end
```

(5) Include your plot here:

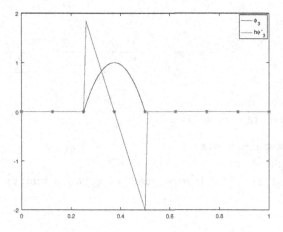

Does ϕ_3' appear to be the derivative of ϕ_3? (<u>yes</u>/no)

(6) Include your plot here:

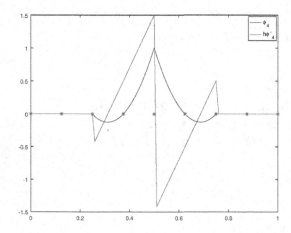

Does ϕ_4' appear to be the derivative of ϕ_4? (<u>yes</u>/no)

(7) For the point $x = 0.4$,

What is the finite difference for ϕ_3? -3.2000

What is the value of phip(3,h,x)? -3.2000

What is the finite difference for ϕ_4? 5.6000

What is the value of phip(4,h,x)? 5.6000

Do they match up? (yes/no)

Solution 13.4.

(1) Include the *final* version of exer4.m, with all additions from this exercise included, here:

```
N=7;
h=1/(N+1);
A1=zeros(N,N);
for m=1:N
  for n=1:N
    A1(m,n)=-gaussquad(@phip,m,@phip,n,h,0,1);
  end
end
A1


% 4.2
ans4_2=norm(A1-A1','fro')


% 4.3
x=linspace(0,1,97);
for k=1:97
  psi(k)=0;
  for n=1:N
    psi(k)=psi(k)+phi(n,h,x(k));
  end
end
plot(x,psi)
print -dpng exer4plot


% 4.4
ans4_4=A1*ones(N,1)


% 4.5
v=(1:N)'*h; % v=x
ans4_5=A1*v


% 4.6
```

```
RHS4=zeros(N,1);
for k=1:N
  RHS4(k)=gaussquad(@rhs4,1,@phi,k,h,0,1);
end
RHS4
xx=(1:N)'*h;
v=xx.*(1-xx);
ans4_6=norm(v-A1\RHS4)
```

(2) Is the matrix A1 symmetric? (<u>yes</u>/no)

(3) Include your plot of ψ here:

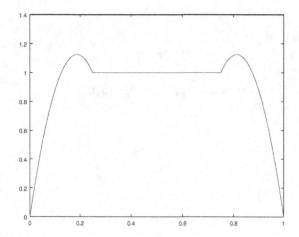

(4) Does A1*ones(N,1) have zeros in positions $m = 3, 4, 5$?
(<u>yes</u>/no)

(5) Do you get a vector that is zero except in the positions $m = 6, 7$?
(<u>yes</u>/no)

(6) What are the values of RHS4?

```
RHS4 =
  -0.33333
  -0.16667
  -0.33333
  -0.16667
  -0.33333
  -0.16667
  -0.33333
```

Do you get the zero vector? (<u>yes</u>/no)

Solution 13.5.

(1) Include the *final* version of `exer5.m`, with all additions from this exercise included, here:

```
% Chapter 13, Exercise 5
% exer5.m
% M. Sussman

N=7;
h=1/(N+1);
A3=zeros(N,N);
for m=1:N
  for n=1:N
    A3(m,n)=gaussquad(@phi,m,@phi,n,h,0,1);
  end
end
A3

% 5.2
ans5_2=norm(A3-A3','fro')

% 5.3

% 5.4
RHS5=zeros(N,1);
for k=1:N
  RHS5(k)=gaussquad(@rhs5,1,@phi,k,h,0,1);
end
RHS5

%5_5
xx=(1:N)'*h;
v=xx.*(1-xx);
ans5_6=norm(v-(A1+A3)\RHS5)
```

(2) Is the matrix `A3` symmetric? (yes/no)

(3) Include a copy of your `rhs5.m` here:

```
function z=rhs5(n,h,x)
  % z=rhs5(n,h,x)

  % M. Sussman

  z=-2+x*(1-x);
end
```

What are the values of RHS5?

```
RHS5 =
  -0.31563
  -0.15052
  -0.29479
  -0.14531
  -0.29479
  -0.15052
  -0.31563
```

(4) Does your solution for U exactly equal $x_n(1 - x_n)$?
(yes/no)

Solution 13.6.

(1) Include the *final* version of **exer6.m**, with all additions from this exercise included, here:

```
N=7;
h=1/(N+1);
A2=zeros(N,N);
for m=1:N
  for n=1:N
    A2(m,n)=gaussquad(@phi,m,@phip,n,h,0,1);
  end
end
A2


% 5.2
ans5_2=norm(A2+A2','fro')


% 5.3
RHS6=zeros(N,1);
for k=1:N
  RHS6(k)=gaussquad(@rhs6,1,@phi,k,h,0,1);
end
RHS6


%5_4
xx=(1:N)'*h;
v=xx.*(1-xx);
ans4_6=norm(v-(A1+A2+A3)\RHS6)
```

(2) Is the matrix A2 skew-symmetric? (yes/no)

(3) Include a copy of your rhs6.m here:

```
function z=rhs6(n,h,x)
  % z=rhs6(n,h,x)

  % M. Sussman

  z=-2+(1-2*x)+x*(1-x);
end
```

What are the values of RHS6?

```
RHS6 =
  -0.19062
  -0.10885
  -0.25312
  -0.14531
  -0.33646
  -0.19219
  -0.44063
```

(4) Does your solution for U exactly equal $x_n(1 - x_n)$? (yes/no)

Solution 13.7.

(1) Include a copy of your exact7.m here:

```
function z=exact7(x)
% z=exact7(x)

  % M. Sussman

  omega=sqrt(3)/2;
  z=x-exp(-(x-1)/2).*sin(omega*x)/sin(omega);
end
```

(2) Include a copy of your rhs7.m here:

```
    @(k,h,x) x+1
```

(3) Include a copy of your solve7.m here:

```
function [x,U]=solve7(N)
  % [x,U]=solve7(N)
  % solve u''+u'+u=x+1  u(0)=u(1)=0 using finite elements
  % with quadratic shape functions
```

```
% M. Sussman

h=1/(N+1);
x=(1:N)'*h;
A1=zeros(N,N);
A2=zeros(N,N);
A3=zeros(N,N);
RHS=zeros(N,1);
for m=1:N
  x0=max(0,(m-2)*h);
  x1=min(1,(m+2)*h);
  for n=max(1,m-2):min(N,m+2)
    A1(m,n)=-gaussquad(@phip,m,@phip,n,h,x0,x1);
    A2(m,n)=gaussquad(@phi,m,@phip,n,h,x0,x1);
    A3(m,n)=gaussquad(@phi,m,@phi,n,h,x0,x1);
  end
  RHS(m)=gaussquad(@(k,h,x) x+1,1,@phi,m,h,x0,x1);
end

U=(A1+A2+A3)\RHS;

end
```

(4) Fill in the following table:

N	h	error	ratio
7	1.2500e-1	7.2202e-06	15.460
15	6.2500e-2	4.6703e-07	15.788
31	3.1250e-2	2.9581e-08	14.013
61	1.6129e-2	2.1110e-09	14.952
121	8.1967e-3	1.4119e-10	—

What is your estimated rate of convergence? $O(h^4)$

Solution 13.8.

(1) Include a copy of your solve8.m here:

```
function [x,U]=solve8(N)
% [x,U]=solve8(N)
% solve u''+u'+u=x+1 u(0)=u(1)=0 using finite elements
% with quadratic shape functions
% Neumann b.c. at x=1

% M. Sussman
```

```
h=1/(N+1);
x=(1:N+1)'*h;
A1=zeros(N+1,N+1);
A2=zeros(N+1,N+1);
A3=zeros(N+1,N+1);
RHS=zeros(N+1,1);
for m=1:N+1
  x0=max(0,(m-2)*h);
  x1=min(1,(m+2)*h);
  for n=max(1,m-2):min(N+1,m+2)
    A1(m,n)=-gaussquad(@phip,m,@phip,n,h,x0,x1);
    A2(m,n)=gaussquad(@phi,m,@phip,n,h,x0,x1);
    A3(m,n)=gaussquad(@phi,m,@phi,n,h,x0,x1);
  end
  RHS(m)=gaussquad(@(k,h,x) x+1,1,@phi,m,h,x0,x1);
end

U=(A1+A2+A3)\RHS;

end
```

(2) Include your plot here:

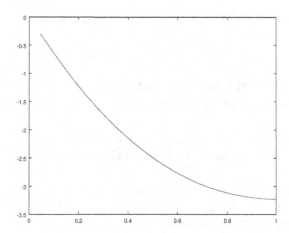

Does $u'(1)$ appear correct? (yes/no)

(3) For the case $N = 21$, what is the value of $u'(1)$?

```
U(20)*phip(20,h,1)+U(21)*phip(21,h,1)+U(22)*phip(22,h,1)
= 0.0029097
```

(4) What is the maximum absolute value of the difference between your computed solution (for $N = 21$) and the exact solution at the nodes?

```
norm(U-exact8(x),inf) =  1.1766e-06
```

Solution 13.9.

(1) Include a copy of your `burgers_ode.m` here:

```
function [F,J]=burgers_ode(t,U)
  % [F,J]=burgers_ode(t,U)
  % compute the right side of the time-dependent ODE arising
  % from a method of lines reduction of Burgers' equation
  % and its derivative, J.
  % Boundary conditions are fixed =0 at the
  % endpoints x=0 and x=1.
  % A fixed number of spatial points (=N) is used.
  % The variable t is not used, but is kept as a place holder.
  % U is the vector of the approximate solution
  % output F is the time derivative of U (column vector)
  % output J is the Jacobian matrix of
  %   partial derivatives of F with respect to U

  % M. Sussman

  % spatial intervals
  N=500;
  NU=0.001;
  dx=1/(N+1);
  ULeft=1;  % left boundary value
  URight=0; % right boundary value

  F=zeros(N,1);  % force F to be a column vector
  J=zeros(N,N);  % matrix of partial derivatives

  % construct F and J in a loop
  for n=1:N
    if n==1  % left boundary
        F(n)   = NU*(U(n+1)-2*U(n)+ULeft )/dx^2- ...
                    U(n)*(U(n+1)-ULeft )/(2*dx);
        J(n,n   )= -2*NU/dx^2-(U(n+1)-ULeft )/(2*dx);
        J(n,n+1)=    NU/dx^2-U(n)/(2*dx);
      elseif n<N  % interior of interval
        F(n)   = NU*(U(n+1)-2*U(n)+U(n-1))/dx^2- ...
```

```
                 U(n)*(U(n+1)-U(n-1))/(2*dx);
      J(n,n-1)=    NU/dx^2+U(n)/(2*dx);
      J(n,n  )= -2*NU/dx^2-(U(n+1)-U(n-1))/(2*dx);
      J(n,n+1)=    NU/dx^2-U(n)/(2*dx);
    else % right boundary
      F(n)   = NU*(URight-2*U(n)+U(n-1))/dx^2- ...
                 U(n)*(URight-U(n-1))/(2*dx);
      J(n,n-1)=    NU/dx^2+U(n)/(2*dx);
      J(n,n  )= -2*NU/dx^2-(URight-U(n-1))/(2*dx);
    end
  end
end
```

(2) Nothing required.

(3) Do these values agree with the given values? (yes/no)

(4) What is the value of U(200,50) (Use `format long.`)?
    0.979077601035924

(5) Include your plot for k=50 here:

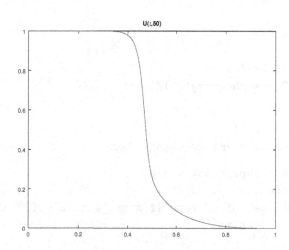

(6) Include the final frame from your "flicker picture" (k=100) here:

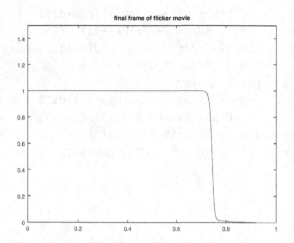

final frame of flicker movie

Solution 13.10.

(1) Include a copy of your `rope_ode.m` here:

```
function fValue=rope_ode(x,y)
  % fValue=rope_ode(x,y) computes the
  % rhs of the first-order system
  % your name and the date

  % M. Sussman

  rhog=0.4;
  c=0.05;
  fValue(1,1)=y(2);
  fValue(2,1)=(rhog-c*y(2))/(1+c*x);
end
```

(2) What value of α? -1

(3) Include a copy of your `rope_shoot.m` here:

```
function F = rope_shoot ( alpha )
  % F = rope_shoot ( alpha )
  % compute height of rope if start with u'=y(2)=alpha
  %    and end at x=5

  % M. Sussman

  [x,y]=ode45(@rope_ode,[0,5],[1;alpha]);
  F=y(end,1)-1.5;
end
```

(4) Does `rope_shoot` return the same value you obtained above when `alpha=0`?
(yes/no)

(5) What is the value of `alpha` you found? -0.850805

(6) Include your plot here:

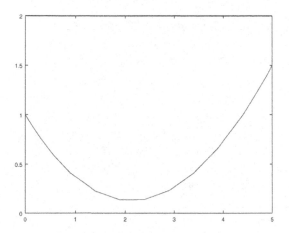

Does the curve have a height of 1 at $x = 0$ and a height of 1.5 at $x = 5$?
(yes/no)

(7) What is your estimate of the derivative of `U2` at the left endpoint?
`(U2(2)-U2(1))/(x2(2)-x2(1))=-0.841597`

How does it compare with the value α you just computed? Agree to many decimal places.

Chapter 14

Vectors, matrices, norms and errors

Solution 14.1. Fill in the following table:

	L^1	L^2	L^∞
x1	17	10.050	7
x2	18	10.488	7
x3	10	6.4807	5

Solution 14.2. Fill in the following table:

Matrix norm (p)	Vector norm (q)	S/U	A=A1 x=x1	A=A1 x=x2	A=A1 x=x3	A=A2 x=x1	A=A2 x=x2	A=A2 x=x3
1	1	S	<=	<=	<=	<=	<=	<=
1	2	U	<=	<=	<=	<=	>	<=
1	inf	U	<=	<=	<=	>	>	>
2	1	U	>	<=	>	<=	<=	<=
2	2	S	<=	<=	<=	<=	<=	<=
2	inf	U	<=	<=	<=	>	>	>
inf	1	U	>	<=	>	<=	<=	<=
inf	2	U	>	<=	<=	<=	<=	<=
inf	inf	S	<=	<=	<=	<=	<=	<=
'fro'	1	U	<=	<=	>	<=	<=	<=
'fro'	2	S	<=	<=	<=	<=	<=	<=
'fro'	inf	U	<=	<=	<=	>	>	>

If you wrote a script file, include it here:

```
% script file for Chapter 14, Exercise 2

x1 = [ 4; 6; 7 ];
x2 = [ 7; 5; 6 ];
x3 = [ 1; 5; 4 ];
```

```
A1 = [38    37    80
       53    49    49
       23    85    46];

A2 = [77    89    78
        6    34    10
       65    36    26];

matnrms={1,2,inf,'fro'};
vecnrms={1,2,inf};
As={A1,A2};
xs={x1,x2,x3};

clear r
for p=1:4
  for q=1:3
     for i=1:2
        for j=1:3
           r=norm(As{i}*xs{j},vecnrms{q})/(norm(As{i},matnrms{p}) ...
                                     *norm(xs{j},vecnrms{q}));
           if r >= 1
             fprintf('A%0.1d,x%0.1d, matnrm %d, vecnrm',i,j,p)
             fprintf('%d, >= r=%e\n',vecnrms{q},r)
           else
             fprintf('A%0.1d,x%0.1d, matnrm %d, vecnrm',i,j,p)
             fprintf('%d, < r=%e\n',vecnrms{q},r)
           end
        end
     end
     fprintf('\n')
  end
end
```

Solution 14.3.

(2) How long does it take for **norm(A)**? 0.486 seconds
(3) How long does it take for the Frobenius norm of A? 0.0442 seconds
(4) Which takes longer, the two norm or Frobenius norm? Frobenius

Solution 14.4.

(1) What are the eigenvalues? 0.5 (one eigenvalue)
    How many linearly independent eigenvectors are there? 1 (one eigenvector)
(2) What is the spectral radius of A? 0.5

(3) What is $\|\mathbf{x}\|_2$? 2.6458

What is $\|A\mathbf{x}\|_2$? 3.7081

What is $(\rho(A))(\|\mathbf{x}\|_2)$? 1.3229

(4) $\|Ax\|_2 \le \rho(A)\|x\|_2$? (yes/<u>no</u>)

Solution 14.5.

(1) Include your plot here:

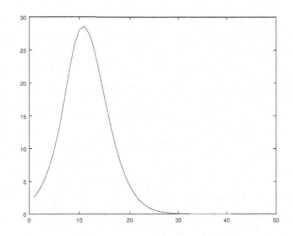

(2) What is the smallest value of $k > 2$ for which $\|A^k x\|_2 \le \|Ax\|_2$? 20

(3) What is $\max_{0 \le k \le 40} \|A^k x\|_2$? 28.597

Solution 14.6.

Fill in the following table:

Case	Residual	L/S	xTrue	Error	L/S
1	9.9998e-13	S	[0;0]	1.4142	L
2	1.4142e-05	S	[1;0]	1.0000e-05	S
3	281.43	L	[1;0]	140.72	L
4	1.0000e+09	L	[1;1]	1.0000e-03	S

Solution 14.7.

(2) Include your plot here:

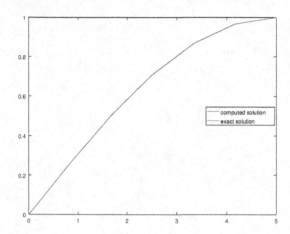

Are the two curves close? (yes/no)

(3) Fill in the following table:

Euclidean (L^2) vector norm

	error ratio	relative error ratio
10/ 20	2.6397	3.5741
20/ 40	2.7286	3.7701
40/ 80	2.7770	3.8802
80/160	2.8023	3.9388
160/320	2.8153	3.9691
320/640	2.8218	3.9845

(4) Repeat the above for the L^1 vector norm.

(L^1) vector norm

	error ratio	relative error ratio
10/ 20	1.8995	3.5152
20/ 40	1.9497	3.7408
40/ 80	1.9749	3.8657
80/160	1.9875	3.9316
160/320	1.9937	3.9655
320/640	1.9969	3.9827

(5) Repeat the above for the L^∞ vector norm.

(L^∞) vector norm

	error ratio	relative error ratio
10/ 20	3.6455	3.6455
20/ 40	3.8106	3.8106
40/ 80	3.9037	3.9037
80/160	3.9504	3.9504
160/320	3.9752	3.9752
320/640	3.9876	3.9876

(6) Fill in the following table with the words "error", "relative error", or "both".

	Rate= $O(h^2)$?
L^2	relative error
L^1	relative error
L^∞	both

Solution 14.8.

Fill in the following table

Results using a recent version of octave

Matrix size	cond(A)	eps*cond(A)	‖difference‖/‖x‖
6	4208.3	9.3442e-13	2.3828e-13
12	4.2674e+09	9.4754e-07	2.1077e-07
18	8.1168e+16	18.023	0.24366
24	4.5571e+17	101.19	0.24366

Results using an older version of MATLAB

Matrix size	cond(A)	eps*cond(A)	‖difference‖/‖x‖
6	5.8650e+003	9.3442e-013	3.6582e-014
12	5.6817e+009	9.4754e-007	3.1622e-008
18	9.7742e+016	19.3423	0.3760
24	4.5002e+018	173.5271	29.4226

Solution 14.9.

(1) Include your calculation here: For $j > 1$ and $j < N$, and writing $\omega = j\pi/(N+1)$,

$$Av^{(j)} = -\sin((j-1)\omega) + 2\sin(j\omega) - \sin((j+1)\omega)$$

But

$$-\sin((j-1)\omega) - \sin((j+1)\omega) =$$
$$-\sin(j\omega)\cos(\omega) + \cos(j\omega)\sin(\omega) - \sin(j\omega)\cos(\omega) - \cos(j\omega)\sin(\omega)$$
$$= -2\sin(j\omega)\cos(\omega)$$

so that
$$Av^{(j)} = 2(1 - \cos(\omega))\sin(j\omega)$$
and hence
$$Av^{(j)} = 2(1 - \cos(\omega))v^j = \lambda_j v^{(j)}$$
with $\lambda_j = 2(1 - \cos(j\pi/(N+1)))$. The same calculation works for $j = 1$ because $\sin((j-1)\omega) = 0$ and for $j = N$, $\sin((j+1)\omega) = 0$.

(2) Include your code for `tridiagonal.m` here:

```
tridiagonal=@(N) 2*diag( ones(N,1) ) - ...
              diag( ones(N-1,1) ,1) - diag( ones(N-1,1) ,-1);
```

(3) Does your code check against (14.4) for N=5? (yes/no)
(4) What is the result for N=25 and j=1? 1.1870e-15
   For j=10? 8.7139e-15

```
A=tridiagonal(25);
N=25;k=(1:N)';
j=1;v=sin(j*pi*k/(N+1));lambda=2*(1-cos(j*pi/(N+1)));
norm(A*v-lambda*v)
ans =    1.1870e-15
j=10;v=sin(j*pi*k/(N+1));lambda=2*(1-cos(j*pi/(N+1)));
norm(A*v-lambda*v)
ans =    8.7139e-15
```

Solution 14.10.

(1) Include a copy of your `rrs.m` here:

```
function A=rrs(A,j)
  % A=rrs(A,j) performs one row reduction step for row j
  % j must be between 2 and N

  % M. Sussman
  N=size(A,1);
  if (j<=1 | j>N)  % | means "or"
     j
     error('rrs: value of j must be between 2 and N');
  end

  factor=1/A(j-1,j-1);  % factor to multiply row (j-1) by

  %<< include code to multiply row (j-1) by factor and >>
  %<< add it to row (j).  Row (j-1) remains unchanged. >>
  A(j,:)=A(j,:)+factor*A(j-1,:);
end
```

Does your work check out? (yes/no)

(2) What is the determinant? 6
(3) What is the determinant after one step? 6
(4) Include a copy of your script file here:

```
tridiagonal=@(N) 2*diag( ones(N,1) ) - ...
                diag( ones(N-1,1) ,1) - diag( ones(N-1,1) ,-1);


N=25;
A=tridiagonal(N);

dets=0;
for k=2:N
  A=rrs(A,k);
   dets(k)=det(A);
end

vals=(2:N+1)'./(1:N)';
d=diag(A);

%are all diagonals correct?
norm(vals-d)

%are all determinants the same?
max(dets(2:end))-min(dets(2:end))

%are product of diagonals and determinant the same?
prod(d)-det(A)
```

Has the determinant changed? (yes/<u>no</u>)
(5) Does the product of the diagonals equal the determinant? (<u>yes</u>/no)

Solution 14.11.

(1) Include your recursive function here:

```
function d=rdet(A)
  % d=rdet(A)
  % recursive function to compute the determinant of A by the
  % Laplace rule (expansion by minors)

  % M. Sussman

  [m,n]=size(A);
  if m~=n
    error('rdet: A is not square');
```

```
elseif n==0
  error('rdet: A is empty');
end

if m==1
  d=A(1,1);
else
  % expand down first column
  d=0;
  for k=1:n
    d=d-(-1)^k*A(k,1)*rdet(A([1:k-1,k+1:n],2:n));
  end
end
end
```

Does your function work properly for A=magic(5)? (<u>yes</u>/no)

(2) How long does magic(7) take? 0.209

How long does magic(8) take? 1.634

How long does magic(9) take? 14.42

How long does magic(10) take? 144.43

Does your timing confirm the $O(N!)$ estimate? (<u>yes</u>/no)

Chapter 15

Solving linear systems

Solution 15.1.

(1) Include a copy of your exer1.m) here:

```
function elapsedTime=exer1(n)
  % elapsedTime=exer1(n)
  % computes elapsed time for computing the inverse
  % of a n X n "magic" matrix and the solution
  % to the system with a vector of ones on the right side

  % M. Sussman

  if mod(n,2)==0
    error('Please use only odd values for n');
  end

  A = magic(n);        % only odd n yield invertible matrices
  b = ones(n,1); % the right side vector doesn't change the time
  tic;
  Ainv = inv(A);       % compute the inverse matrix
  xSolution = Ainv*b;  % compute the solution
  elapsedTime=toc;
end
```

(2) Fill in the following table:

Time to compute inverse matrices		
n	time	ratio
161	0.0037	2.1920
321	0.0080	6.6466
641	0.0535	6.7805
1281	0.3627	7.4577
2561	2.7046	7.4786
5121	20.2263	7.6735
10241	155.2077	—

(3) Are these solution times roughly proportional to n^3? (yes/no)

Solution 15.2.

(1) Include a copy of your `exer2.m` here:

```
function elapsedTime=exer2(n)
  % elapsedTime=exer2(n)
  % computes elapsed time for computing the solution
  % of a n X n "magic" matrix in
  % the system with a vector of ones on the right side

  % M. Sussman

  if mod(n,2)==0
    error('Please use only odd values for n');
  end

  A = magic(n);      % only odd n yield invertible matrices
  b = ones(n,1); % the right side vector doesn't change the time
  tic;
  xSolution = A\b;  % compute the solution
  elapsedTime=toc;
end
```

(2) Fill in the following table:

Time to compute solutions		
n	time	ratio
161	0.0021	1.8561
321	0.0038	6.1150
641	0.0235	6.0935
1281	0.1432	6.9328
2561	0.9930	7.0661
5121	7.0167	7.5437
10241	52.9325	—

(3) Are these solution times roughly proportional to n^3? (yes/no)

(4) Fill in the following table:

Comparison of times	
n	(time for inverse)/(time for solution)
161	1.7755
321	2.1332
641	2.2576
1281	2.5140
2561	2.7141
5121	2.8708
10241	2.9334

(5) Are your results consistent with theory? (yes/no)

Solution 15.3.

(1) Is the relative norm of the difference zero or roundoff? (yes/no)

(2) Fill in the following table:

Time to compute sparse matrix solutions		
n	time	ratio
10240	0.0011	2.45
102400	0.0027	11.3
1024000	0.0307	9.8
10240000	0.3017	—

(3) What is your estimate of p? 1

(4) What is the comparison? For size 10240, sparse time $= 0.0011$, full time $= 52$. Sparse is hugely faster.

Solution 15.4.

(1) For the matrix defined by **tridiagonal**, fill in the following table:

tridiagonal matrix			
size	error	determinant	cond. no.
10	6.5460e-16	11.000	48.374
40	4.1311e-15	41.000	680.62
160	9.5640e-15	161.00	1.0505e+04

(2) For the matrix defined by `hilb`, fill in the following table:

Hilbert matrix			
size	error	determinant	cond. no.
10	1.9144e-04	2.1644e-53	1.6025e+13
15	0.0010124	-2.1903e-120	1.1054e+18
20	0.059753	-1.1004e-195	2.6383e+18

(3) For the matrix defined by `frank`, fill in the following table:

Frank matrix			
size	error	determinant	cond. no.
10	3.3894e-11	1.00000	2.8543e+07
15	1.3640e-04	0.99999	1.3710e+13
20	0.077073	9.2784	8.6605e+17

(4) For the matrix defined by `pascal`, fill in the following table:

Pascal matrix			
size	error	determinant	cond. no.
10	1.1555e-08	1	4.1552e+09
15	0.0035304	1	2.8396e+15
20	0.011764	-15.914	1.2281e+20

Solution 15.5.

(1) What is the matrix U?

```
U =
    1.00000   0.50000   0.33333   0.25000   0.20000
    0.00000   0.08333   0.08333   0.07500   0.06667
    0.00000   0.08333   0.08889   0.08333   0.07619
    0.00000   0.07500   0.08333   0.08036   0.07500
    0.00000   0.06667   0.07619   0.07500   0.07111
```

Are all entries in the first column in the second through last rows are zero or roundoff? (yes/no)

(2) What is `norm(L*U-A,'fro')/norm(A,'fro')`? 0

(3) Include a copy of your `gauss_lu.m` here:

```
function [L,U]=gauss_lu(A)
  % [L,U]=gauss_lu(A) performs an LU factorization of
  % the matrix A using Gaussian reduction.
  % A is the matrix to be factored.
  % L is a lowere triangular factor with 1's on the diagonal
  % U is an upper triangular factor.
  % A = L * U

  % M. Sussman

  [n,m]=size(A);
  if n ~= m
    error('gauss_lu: A is not a square matrix.')
  end

  L=eye(n);      % square n by n identity matrix
  U=A;
  for Jcol=1:n
    for Irow=Jcol+1:n
      % compute Irow multiplier and save in L(Irow,Jcol)
      L(Irow,Jcol)=U(Irow,Jcol)/U(Jcol,Jcol);

      % multiply row "Jcol" by L(Irow,Jcol)
      %    and subtract from row "Irow"
      % This vector statement could be replaced with a loop
      U(Irow,Jcol:n)=U(Irow,Jcol:n)-L(Irow,Jcol)*U(Jcol,Jcol:n);
    end
  end
end
```

(4) (nothing required)

(5) U(5,5), to at least four significant figures is 2.2676e-05.

(6) Is L lower triangular? (yes/no)

   Is U upper triangular? (yes/no)

(7) Does the product L*U recover the original matrix? (yes/no)

(8) (a) Use norms to confirm that LR*UR=R, without printing the matrices. Include the commands you used.

```
   norm(R-LR*UR,'fro')/norm(R,'fro')
   ans =     4.3740e-14
```

 (b) Use MATLAB functions `tril` and `triu` to confirm that LR is lower triangular and UR is upper triangular, without printing the matrices.

```
   norm(UR-triu(UR),'fro')/norm(UR,'fro')
```

```
ans =      1.8298e-17
norm(LR-tril(LR),'fro')/norm(LR,'fro')
ans = 0
```

Solution 15.6.

(1) What are L1 and U1?

```
L1 =
    1.00000   0.00000   0.00000   0.00000   0.00000
   -0.50000   1.00000   0.00000   0.00000   0.00000
   -0.00000   0.00000   1.00000   0.00000   0.00000
   -0.50000  -3.00000       Inf   1.00000   0.00000
   -0.00000   2.00000      -Inf       NaN   1.00000


U1 =
   -2.00000   1.00000   0.00000   0.00000   0.00000
    0.00000   0.50000   1.00000  -2.00000   0.00000
    0.00000   0.00000   0.00000   1.00000  -2.00000
    0.00000   0.00000       NaN      -Inf       Inf
    0.00000   0.00000       NaN       NaN       NaN
```

(2) On which step of the decomposition (values of Irow and Jcol) does the method
    fail? Jcol=3, Irow=4
    Why does it fail? A1(3,3) is zero, fails on division by zero (gives Inf)
(3) What is the determinant of A1? -8
    What is the condition number (cond(A1))? 15.4
    Is the matrix singular or ill-conditioned? (yes/<u>no</u>)

Solution 15.7.

(1) What is A2=P1*A?

```
A2 =
   16    2    3   13
    9    7    6   12
    5   11   10    8
    4   14   15    1
```

(2) What is A3=A*P1?

```
A3 =
   16    3    2   13
    5   10   11    8
    9    6    7   12
    4   15   14    1
```

(3) What permutation matrix would interchange the first and fourth rows of A?

```
[ 0 0 0 1
  0 1 0 0
  0 0 1 0
  1 0 0 0 ]
```

(4) What is the product P=P1*P2?

```
P =
    1   0   0   0
    0   0   1   0
    0   0   0   1
    0   1   0   0
```

(5) What is the product A4=P*A?

```
A4 =
   16    2    3   13
    9    7    6   12
    4   14   15    1
    5   11   10    8
```

Solution 15.8.

(1) Include your *final* gauss_plu.m here:

```
function [P,L,U]=gauss_plu(A)
  % [P,L,U]=gauss_lu(A) performs a PLU factorization of
  % the matrix A using Gaussian reduction.
  % A is the matrix to be factored.
  % P is a permutation matrix
  % L is a lower triangular factor with 1's on the diagonal
  % U is an upper triangular factor.
  % A = P * L * U

  % M. Sussman

  [n,m]=size(A);
  if n ~= m
    error('gauss_lu: A is not a square matrix.')
  end

  L=eye(n);      % square n by n identity matrix
  P=L;
  U=A;
```

```
    for Jcol=1:n
      % First, choose a pivot row by finding the largest entry
      % in column=Jcol on or below position k=Jcol.
      % The row containing the largest entry will then be switched
      % with the current row=Jcol
      [ colMax, pivotShifted ] = max ( abs ( U(Jcol:n, Jcol ) ) );

      % The value of pivotShifted from max needs to be adjusted.
      % See help max for more information.
      pivotRow = Jcol+pivotShifted-1;

      if pivotRow ~= Jcol      % no pivoting if pivotRow==Jcol
        U([Jcol, pivotRow], :)        = U([pivotRow, Jcol], :);
        L([Jcol, pivotRow], 1:Jcol-1)= L([pivotRow, Jcol], 1:Jcol-1);
        P(:,[Jcol, pivotRow])         = P(:,[pivotRow, Jcol]);
      end
      for Irow=Jcol+1:n
        % compute Irow multiplier and save in L(Irow,Jcol)
        L(Irow,Jcol)=U(Irow,Jcol)/U(Jcol,Jcol);

        % multiply row "Jcol" by L(Irow,Jcol) and
        %    subtract from row "Irow"
        % This vector statement could be replaced with a loop
        U(Irow,Jcol:n)=U(Irow,Jcol:n)-L(Irow,Jcol)*U(Jcol,Jcol:n);
      end
      %{
      % TEMPORARY DEBUGGING CHECK
      if norm(P*L*U-A,'fro')> 1.e-12 * norm(A,'fro')
        error('If you see this message, there is a bug!')
      end
      %}
    end
  end
```

(2) What are P, L and U?

```
P =
    1   0   0   0   0
    0   0   0   0   1
    0   0   1   0   0
    0   0   0   1   0
    0   1   0   0   0
```

```
L =
    1.00000    0.00000    0.00000    0.00000    0.00000
    1.00000    1.00000    0.00000    0.00000    0.00000
    1.00000    0.50000    1.00000    0.00000    0.00000
    1.00000    0.75000    0.75000    1.00000    0.00000
    1.00000    0.25000    0.75000   -1.00000    1.00000

U =
    1.00000    1.00000    1.00000    1.00000    1.00000
    0.00000    4.00000   14.00000   34.00000   69.00000
    0.00000    0.00000   -2.00000   -8.00000  -20.50000
    0.00000    0.00000    0.00000   -0.50000   -2.37500
    0.00000    0.00000    0.00000    0.00000   -0.25000
```

Is P*L*U=A? (<u>yes</u>/no)
Is P the identity matrix? (yes/<u>no</u>)

(3) What are P1, L1 and U1?

```
P1 =
    1    0    0    0    0
    0    0    0    1    0
    0    0    0    0    1
    0    1    0    0    0
    0    0    1    0    0

L1 =
    1.00000    0.00000    0.00000    0.00000    0.00000
   -0.50000    1.00000    0.00000    0.00000    0.00000
   -0.00000   -0.66667    1.00000    0.00000    0.00000
   -0.50000   -0.33333   -1.00000    1.00000    0.00000
   -0.00000   -0.00000   -0.00000   -1.00000    1.00000

U1 =
   -2.00000    1.00000    0.00000    0.00000    0.00000
    0.00000   -1.50000    1.00000    0.00000    0.00000
    0.00000    0.00000   -1.33333    1.00000    0.00000
    0.00000    0.00000    0.00000   -1.00000    0.00000
    0.00000    0.00000    0.00000    0.00000   -2.00000
```

Were there any error messages or NaN or inf? (yes/<u>no</u>)

(4) Nothing required.

Solution 15.9.

(1) Is $A = PLU$? (<u>yes</u>/no)

(2) Fill in:

Step 0	Step 1	Step 2	Step 3
b	z	y	x
$\begin{bmatrix} 28 \\ 18 \\ 16 \end{bmatrix}$	$\begin{bmatrix} 16 \\ 28 \\ 18 \end{bmatrix}$	$\begin{bmatrix} 16 \\ 20 \\ 4 \end{bmatrix}$	$\begin{bmatrix} -1 \\ 1 \\ 2 \end{bmatrix}$

Solution 15.10.

(1) Include your p_solve.m here:

```
function z=p_solve(P,b)
  % z=p_solve(P,b)
  % P is an orthogonal matrix
  % b is the right hand side
  % z is the solution of P*z=b

  % M. Sussman

  z=P'*b;
end
```

(2) Does your result agree with Step 1? (<u>yes</u>/no)

(3) Include your l_solve.m here:

```
function y=l_solve(L,z)
  % y=l_solve(L,z)
  % L is a lower-triangular matrix whose diagonal entries are 1
  % z is the right hand side
  % y is the solution of L*y=z

  % M. Sussman

  % set n for convenience and simplicity
  n=numel(z);
  % initialize y to zero and make sure it is a column vector
  y=zeros(n,1);

  % first row is really an easy one, especially since the diagonal
  % entries of L are equal to 1
  Irow=1;
  y(Irow)=z(Irow);

  for Irow=2:n
```

```
      rhs = z(Irow);
      for Jcol=1:Irow-1
        rhs = rhs - L(Irow,Jcol)*y(Jcol);
      end
      y(Irow) = rhs;
    end
  end
```

(4) Does your result agree with Step 2? (yes/no)
(5) Include your u_solve.m here:

```
function x = u_solve(U,y)
  % x=u_solve(U,y)
  % U is an upper-triangular matrix
  % y is the right hand side
  % x is the solution of U*x=y

  % M. Sussman

  % set n for convenience and simplicity
  n=numel(y);
  % initialize y to zero and make sure it is a column vector
  x=zeros(n,1);

  % last row is the easy one
  Irow=n;
  x(Irow)=y(Irow)/U(Irow,Irow);

  % the -1 in the middle means it is going up from the bottom
  for Irow=n-1:-1:1
    rhs = y(Irow);
    % the following loop could also be written as a single
    % vector statement.
    for Jcol=Irow+1:n
      rhs = rhs - U(Irow,Jcol)*x(Jcol);
    end
    x(Irow) = rhs/U(Irow,Irow);
  end
end
```

(6) Does your result agree with Step 3? (yes/no)
(7) What is x?

```
x =
   -1
    1
    2
```

Is relErr = norm(P*L*U*x-b)/norm(b) zero or roundoff? (yes/no)

(8) What is relErr? 4.4352e-14

Solution 15.11.

(2) What is CalculationTime? 104.1853
Save your solution: (usave=u).

(3) Include your bels1.m here:

```
% Chapter 15, exercise 11
% File named l_bels1.m
% M. Sussman

% ntimes = number of temporal steps from t=0 to t=1 .
ntimes=30;
% dt = time increment
dt=1/ntimes;
t(1,1)=0;

% N is the dimension of the space
N=402;

% initial values
u(1:N/3   ,1) = linspace(0,1,N/3)';
u(N/3+1:2*N/3,1) = linspace(1,-1,N/3)';
u(2*N/3+1:N,1) = linspace(-1,0,N/3)';

% discretization matrix is -tridiagonal multiplied
% by a constant to make the solution interesting
A= -N^2/5 * tridiagonal(N);
EulerMatrix=eye(N)-dt*A;

tic;
[P,L,U] = gauss_plu(EulerMatrix);
for k = 1 : ntimes
  t(1,k+1) = t(1,k) + dt;
  u(:,k+1) = plu_solve(P,L,U,u(:,k));
end
CalculationTime=toc
```

```
% plot the solution at sequence of times
plot(u(:,1),'r')
hold on
for k=2:3:ntimes
  plot(u(:,k))
end
plot(zeros(size(u(:,1))),'g')
title 'Solution at selected times'
hold off
```

Are your solutions the same in about the same time? (yes/no)
(4) How long does bels1.m take? 3.8601
(5) What is norm(usave-u,'fro')? 0

Solution 15.12.

(1) Does it have only a single interchange of rows? (yes/no)
(2) Include your partial new_plu.m here:

```
function [P,L,U]=new_plu(A)
% [P,L,U]=new_lu(A) performs a PLU factorization of
% the matrix A using Gaussian reduction.
% A is the matrix to be factored.
% P is a permutation matrix
% L is a lower triangular factor with 1's on the diagonal
% U is an upper triangular factor.
% A = P * L * U

% M. Sussman

[n,m]=size(A);
if n ~= m
  error('gauss_lu: A is not a square matrix.')
end

L=eye(n);     % square n by n identity matrix
P=L;
U=A;
%for Jcol=1:n
for Jcol=1:3
  % First, choose a pivot row by finding the largest entry
  % in column=Jcol on or below position k=Jcol.
  % The row containing the largest entry will then be switched
```

```
% with the current row=Jcol
[ colMax, pivotShifted ] = max ( abs ( U(Jcol:n, Jcol ) ) );

% The value of pivotShifted from max needs to be adjusted.
% See help max for more information.
pivotRow = Jcol+pivotShifted-1;

if pivotRow ~= Jcol      % no pivoting if pivotRow==Jcol
  U([Jcol, pivotRow], :)      = U([pivotRow, Jcol], :);
  L([Jcol, pivotRow], 1:Jcol-1)= L([pivotRow, Jcol], 1:Jcol-1);
  P(:,[Jcol, pivotRow])      = P(:,[pivotRow, Jcol]);
end
for Irow=Jcol+1:n
  % compute Irow multiplier and save in L(Irow,Jcol)
  L(Irow,Jcol)=U(Irow,Jcol)/U(Jcol,Jcol);

  % multiply row "Jcol" by L(Irow,Jcol) and
  %    subtract from row "Irow"
  % This vector statement could be replaced with a loop
  U(Irow,Jcol:n)=U(Irow,Jcol:n)-L(Irow,Jcol)*U(Jcol,Jcol:n);
end
%{
% TEMPORARY DEBUGGING CHECK
if norm(P*L*U-A,'fro')> 1.e-12 * norm(A,'fro')
  error('If you see this message, there is a bug!')
end
%}
  end
end
```

(3) What is the matrix Π?

```
Pi =
   1   0   0   0   0   0
   0   1   0   0   0   0
   0   0   1   0   0   0
   0   0   0   0   0   1
   0   0   0   0   1   0
   0   0   0   1   0   0
```

Is $\Lambda\Pi^T = \Lambda$? (yes/no)

```
Lambda=L-eye(6);
norm(Lambda*Pi'-Lambda,'fro')
```

```
ans = 0
```

Why this is an example of a generally true relation?
Call the current column j. Λ has nonzeros only in columns 1 through j, but Π^T differs from the identity only in columns $j+1$ through n. Hence $\Lambda\Pi^T = \Lambda$.

(4) Include your final new_plu.m here:

```
function [P,L,U]=new_plu(A)
  % [P,L,U]=new_lu(A) performs a PLU factorization of
  % the matrix A using Gaussian reduction.
  % A is the matrix to be factored.
  % P is a permutation matrix
  % L is a lower triangular factor with 1's on the diagonal
  % U is an upper triangular factor.
  % A = P * L * U


  % M. Sussman

  [n,m]=size(A);
  if n ~= m
    error('gauss_lu: A is not a square matrix.')
  end


  L=eye(n);     % square n by n identity matrix
  P=L;
  U=A;
  for Jcol=1:n
  %for Jcol=1:3
    % First, choose a pivot row by finding the largest entry
    % in column=Jcol on or below position k=Jcol.
    % The row containing the largest entry will then be switched
    % with the current row=Jcol
    % colMax, pivotShifted ] = max ( abs ( U(Jcol:n, Jcol ) ) );

    % The value of pivotShifted from max needs to be adjusted.
    % See help max for more information.
    %pivotRow = Jcol+pivotShifted-1;
    colMax=0;
    for j=Jcol:n
      if abs(U(j,Jcol)) > colMax
        pivotRow=j;
        colMax=abs(U(j,Jcol));
      end
```

```
    end

    if pivotRow ~= Jcol     % no pivoting if pivotRow==Jcol
      %U([Jcol, pivotRow], :)        = U([pivotRow, Jcol], :);
      %L([Jcol, pivotRow], 1:Jcol-1)= L([pivotRow, Jcol], 1:Jcol-1);
      %P(:,[Jcol, pivotRow])         = P(:,[pivotRow, Jcol]);
      for k=1:n
        tem=U(Jcol,k);
        U(Jcol,k)=U(pivotRow,k);
        U(pivotRow,k)=tem;

        tem=P(k,Jcol);
        P(k,Jcol)=P(k,pivotRow);
        P(k,pivotRow)=tem;
      end
      for k=1:Jcol-1
        tem=L(Jcol,k);
        L(Jcol,k)=L(pivotRow,k);
        L(pivotRow,k)=tem;
      end
    end

    for Irow=Jcol+1:n
      % compute Irow multiplier and save in L(Irow,Jcol)
      L(Irow,Jcol)=U(Irow,Jcol)/U(Jcol,Jcol);

      % multiply row "Jcol" by L(Irow,Jcol) and
      %    subtract from row "Irow"
      % This vector statement could be replaced with a loop
      U(Irow,Jcol:n)=U(Irow,Jcol:n)-L(Irow,Jcol)*U(Jcol,Jcol:n);
    end
    %{
    % TEMPORARY DEBUGGING CHECK
    if norm(P*L*U-A,'fro')> 1.e-12 * norm(A,'fro')
      error('If you see this message, there is a bug!')
    end
    %}
  end
end
```

Explain how you tested that your code is correct.

```
[P,L,U]=gauss_plu(A);
```

```
[P1,L1,U1]=new_plu(A);
norm(P-P1,'fro')
norm(L-L1,'fro')
norm(U-U1,'fro')
```

and got three zeros for the norms. Then, tried the same sequence with
`A=rand(100,100));` and got three more zeros.

Chapter 16

Factorizations

Solution 16.1.

(1) Include a copy of your unstable_gs.m here:

```
function Q=unstable_gs(X)
  % Q=unstable_gx(X) uses unstable Gram-Schmidt to orthonormalize
  % a set of vectors X

  % M. Sussman

  nx=size(X,2); %number of columns of X

  nq=0;
  for k=1:nx
    y=X(:,k);
    for ell=1:nq
      rlk=dot(Q(:,ell),X(:,k));
      y=y-rlk*Q(:,ell);
    end
    rkk=sqrt(dot(y,y));
    if rkk>0
      nq=nq+1;
      Q(:,nq)=y/rkk;
    end
  end
end
```

(2) Did you get the identity matrix? (<u>yes</u>/no)

(3) Include your Q matrix here:

```
Q =
    5.7735e-01    4.0825e-01    7.0711e-01
```

```
5.7735e-01     4.0825e-01   -7.0711e-01
5.7735e-01    -8.1650e-01   -2.3551e-16
```

(4) Did you get the correct Q matrix? (yes/no)
What MATLAB code did you use to verify that the columns of Q have L^2 norm 1, and are pairwise orthogonal?

```
Q'*Q = 3X3 identity
```

(5) Which condition is false?

```
Q*Q' = identity
```

(6) Is Q1 an orthogonal matrix? (yes/no)
Is it close to an orthogonal matrix? (yes/no)

```
norm(Q1*Q1'-eye(10)) =  3.0016 = norm(Q1'*Q1-eye(10)) NOT ZERO
```

(7) What is the 3 × 3 submatrix at the top left of B1?

```
B1(1:3,1:3) =
    1.0000e+00     1.1553e-15   -1.7455e-14
    1.1553e-15     1.0000e+00   -4.6935e-14
   -1.7455e-14    -4.6935e-14    1.0000e+00
```

What is the 3 × 3 submatrix at the lower right of B1?

```
B1(end-2:end,end-2:end) =
    1.00000    0.99997    0.99985
    0.99997    1.00000    0.99996
    0.99985    0.99996    1.00000
```

What is a 3 × 3 submatrix near the middle of B1?

```
B1(4:6,4:6) =
    1.0000e+00    -1.1314e-08     3.6854e-07
   -1.1314e-08     1.0000e+00     1.7132e-05
    3.6854e-07     1.7132e-05     1.0000e+00
```

Solution 16.2.

(1) Include a copy of your modified_gs.m here:

```
function Q=modified_gs(X)
   % Q=modified_gs(X) uses modified Gram-Schmidt to orthonormalize
   % a set of vectors X

   % M. Sussman
```

```
    nx=size(X,2); %number of columns of X

    nq=0;
    for k=1:nx
      y=X(:,k);
      for ell=1:nq
        rlk=dot(Q(:,ell),y);
        y=y-rlk*Q(:,ell);
      end
      rkk=sqrt(dot(y,y));
      if rkk>0
        nq=nq+1;
        Q(:,nq)=y/rkk;
      end
    end
  end
```

(2) What three matrices did you use?

```
A=rand(3);
```

How did you compare them?

```
norm(modified_gs(A)-unstable_gs(A)) = roundoff
```

(3) What is `norm(B1-eye(10),'fro')`? 3.4655
What is `norm(B2-eye(10),'fro')`? 2.5502e-04

Solution 16.3.

(1) Include a copy of your gs_factor.m here:

```
function [Q,R]=gs_factor(A)
  % [Q,R]=gs_factor(A) uses Gram-Schmidt to do QR factorization

  % M. Sussman

  [m,n]=size(A);

  Q=zeros(m,n);
  R=zeros(n);
  for k=1:n
    y=A(:,k);
    for ell=1:n
      R(ell,k)=dot(Q(:,ell),y);
      y=y-R(ell,k)*Q(:,ell);
```

```
    end
    R(k,k)=sqrt(dot(y,y));
    if R(k,k)>0
      Q(:,k)=y/R(k,k);
    else
      error('gs_factor: rkk is zero')
    end
  end
end
```

(2) Are the matrices Q computed by `gs_factor.m` and `modified_gs.m` the same?
(yes/no)
Is $A = QR$? (yes/no)
Is R upper triangular? (yes/no)

(3) • Is it true that $Q^T Q = I$? (yes/no)
• Is it true that $QQ^T = I$? (yes/no)
• Is Q orthogonal? (yes/no)
• Is the matrix R upper triangular? (yes/no)
• Is it true that $A = QR$? (yes/no)

(4) • Is it true that $Q^T Q = I$? (yes/no)
• Is it true that $QQ^T = I$? (yes/no)
• Is Q orthogonal? (yes/no)
• Is the matrix R upper triangular? (yes/no)
• Is it true that $A = QR$? (yes/no)

Solution 16.4.

(1) Include a copy of your `householder.m` here:

```
function H = householder(b, k)
  % H = householder(b, k)
  % Generate Householder matrix so that
  % H*b has zeros in positions (k+1:end)
  % b is a column vector of length > k

  % your name and the date
  % M. Sussman

  n = size(b,1);
  if n<k
    error('householder: k must be smaller than size of b')
  end
  if size(b,2) ~= 1
    error('householder: b must be a column vector');
```

```
   end

   d = b(k:n);
   if norm(d)==0
     H=eye(n);
     return
   end
   if d(1)>=0
     alpha = -norm(d);
   else
     alpha =  norm(d);
   end

   v(1,1)=sqrt(.5*(1-d(1)/alpha));
   p=-alpha*v(1);
   v(2:n-k+1,1)=d(2:n-k+1)/(2*p);
   w=[zeros(k-1,1);v];
   H=eye(n)-2*w*w';

end
```

(2) (a) For k=1

   Is H orthogonal? (yes/no)

   Does have zeros in positions k+1 and below? (yes/no)

   (b) For k=4

   Is H orthogonal? (yes/no)

   Does have zeros in positions k+1 and below? (yes/no)

Solution 16.5.

(2) What are H1 and A1=H1*A?

```
H1 =
  -0.523387  -0.708111  -0.123150  -0.307875  -0.338662
  -0.708111   0.670851  -0.057243  -0.143108  -0.157419
  -0.123150  -0.057243   0.990045  -0.024888  -0.027377
  -0.307875  -0.143108  -0.024888   0.937779  -0.068443
  -0.338662  -0.157419  -0.027377  -0.068443   0.924713


A1 =
  -3.2481e+01  -2.6631e+01  -2.1397e+01  -2.3706e+01  -2.5861e+01
   1.5543e-15  -1.8535e+01  -3.4109e+00  -7.3797e-01  -2.9935e+00
  -5.5511e-16   1.9070e+00   1.1189e+01   1.7437e+01   1.8697e+01
  -3.3307e-16   1.7675e+00   1.4474e+01   1.4592e+01  -5.2580e+00
```

```
    -1.7764e-15    6.7443e+00    2.0021e+01   -5.0486e+00  -8.3855e-02
```

Are there any non-zero (non-roundoff) values below the diagonal in the first column of A1? (yes/<u>no</u>)

(3) What are H2 and A2=H2*A1?

```
H2 =
     1.00000  -0.00000  -0.00000  -0.00000  -0.00000
    -0.00000  -0.93166   0.09586   0.08885   0.33901
    -0.00000   0.09586   0.99524  -0.00441  -0.01682
    -0.00000   0.08885  -0.00441   0.99591  -0.01559
    -0.00000   0.33901  -0.01682  -0.01559   0.94050


A2 =
    -3.2481e+01  -2.6631e+01  -2.1397e+01  -2.3706e+01  -2.5861e+01
    -2.1331e-15   1.9894e+01   1.2323e+01   1.9439e+00   4.0856e+00
    -3.7213e-16  -9.7145e-17   1.0409e+01   1.7304e+01   1.8345e+01
    -1.6347e-16   1.1102e-16   1.3750e+01   1.4469e+01  -5.5836e+00
    -1.1292e-15  -1.7764e-15   1.7260e+01  -5.5193e+00  -1.3262e+00
```

Are there any non-zero (non-roundoff) values below the diagonal in the first and second columns of A2? (yes/<u>no</u>)

Solution 16.6.

(1) Include a copy of your m-file h_factor.m here:

```
function [Q,R]=h_factor(A)
  % function [Q,R]=h_factor(A) computes the QR factorization using
  % Householder matrices

  % M. Sussman

  [m,n]=size(A);

  Q=eye(m);
  R=A;
  for k=1:n
    H=householder(R(:,k),k);
    Q=Q*H';
    R=H*R;
  end
end
```

(2) Is R upper triangular? (<u>yes</u>/no) Is Q equivalent to a permutation matrix? (yes/<u>no</u>) Does Q*R = A? (<u>yes</u>/no)

(3) Do your results compare with those from gs_factor? (<u>yes</u>/no)

(4) Do your results compare with those from qr? (<u>yes</u>/no)

(5) Show that B3=Q3'*Q3 is much closer to the identity matrix than either B2 or B1

```
norm(B1-eye(10),'fro') =    3.4655
norm(B2-eye(10),'fro') =    2.5502e-04
norm(B3-eye(10),'fro') =    1.3863e-15
```

Solution 16.7.

(2) Include a copy of your h_solve.m here:

```
function x=h_solve(Q,R,b)
  %x=h_solve(Q,R,b) solve the system Q*R*x=b

  % M. Sussman

  x=u_solve(R,Q'*b);
end
```

(3) Is norm(x - x2)/norm(x) close to zero? (<u>yes</u>/no)

Solution 16.8.

(1) Why is L1 *not* the Cholesky factor of A?

The matrix L1 has negative numbers on the diagonal.

(2) Include a copy of your cholesky.m here:

```
function L=cholesky(A)
  % L=cholesky(A) computes the lower-triangular Cholesky factor of A

  % M. Sussman

  [m,n]=size(A);
  if m ~= n
    error('cholesky: A is not a square matrix')
  end

  L=zeros(n,n);
  for j=1:n
    elljj=( A(j,j) -sum(L(j,1:j-1).^2));
    if elljj <= 0
      error('cholesky: A is not positive definite');
    else
      L(j,j)=sqrt(elljj);
```

```
      end
      for k=j+1:n
        L(k,j)=(A(k,j)-sum(L(k,1:j-1).*L(j,1:j-1)) )/L(j,j);
      end
    end
  end
```

(3) Is L*L'=A? (yes/no)

(4) Compare the results from `chol` with those from your `cholesky` function on a matrix of your choice.

```
L=cholesky(A)
L =
     1.41421   0.00000   0.00000   0.00000   0.00000
     0.70711   1.58114   0.00000   0.00000   0.00000
     0.70711   1.58114   2.00000   0.00000   0.00000
     0.70711   2.21359   3.00000   2.56905   0.00000
     0.70711   2.84605   5.00000   5.13809   3.31662
```

```
LL=chol(A,'lower')
LL =
     1.41421   0.00000   0.00000   0.00000   0.00000
     0.70711   1.58114   0.00000   0.00000   0.00000
     0.70711   1.58114   2.00000   0.00000   0.00000
     0.70711   2.21359   3.00000   2.56905   0.00000
     0.70711   2.84605   5.00000   5.13809   3.31662
```

```
norm(L-LL,'fro') =    3.5941e-15
```

(5) Include a copy of your modified `l_solve.m` here:

```
function y=l_solve(L,z)
  % y=l_solve(L,z)
  % L is a lower-triangular matrix whose diagonal entries are 1
  % z is the right hand side
  % y is the solution of L*y=z

  % M. Sussman

  % set n for convenience and simplicity
  n=numel(z);
  % initialize y to zero and make sure it is a column vector
  y=zeros(n,1);
```

```
% first row is really an easy one
Irow=1;
y(Irow)=z(Irow)/L(Irow,Irow);

for Irow=2:n
  rhs = z(Irow);
  for Jcol=1:Irow-1
    rhs = rhs - L(Irow,Jcol)*y(Jcol);
  end
  y(Irow) = rhs/L(Irow,Irow);
end
end
```

How did you check it?

```
A=pascal(5,1)
A =
   1   0   0   0   0
   1  -1   0   0   0
   1  -2   1   0   0
   1  -3   3  -1   0
   1  -4   6  -4   1
x=l_solve(A,[1;1;2;3;5])
norm(A*x-[1;1;2;3;5]) = 0
```

(6) Nothing required.

(7) Solve the system $Ax_0 = b$ and show that $\|x_0 - x\|/\|x\|$ is small.

```
N = 50;
A = rand(N,N);          % random numbers between 0 and 1
A = .5 * (A + A');      % force A to be symmetric
A = A + diag(sum(A));   % make A positive definite by Gershgorin
x = rand(N,1);          % solution
b = A*x;                % right side

L=cholesky(A);
y=l_solve(L,b);
x0=u_solve(L',y);
norm(x-x0)/norm(x) = roundoff number
```

Chapter 17

The eigenvalue problem

Solution 17.1.

(1) Include a copy of your `rayleigh.m` here:

```
function r=rayleigh(A,x)
  % r=rayleigh(A,x) computes the Rayleigh quotient
  % returns it in r

  % M. Sussman

  r=x'*A*x/(x'*x);
end
```

(2) Nothing required
(3) Is [1;0;1] and eigenvector with eigenvalue 1? (yes/no)
   Is [0;1;1] and eigenvector with eigenvalue -1.5? (yes/no)
   Is [1;-2;0] and eigenvector with eigenvalue 2? (yes/no)
(4) Compute the value of the Rayleigh quotient for the matrix `A=eigen_test(1)` and vectors in the following table.

x	R(A,x)	
Some Rayleigh quotients		
[3; 2; 1]	4.5	
[1; 0; 1]	1	(is an eigenvector)
[0; 1; 1]	-1.5	(is an eigenvector)
[1;-2; 0]	2	(is an eigenvector)
[1; 1; 1]	3	
[0; 0; 1]	-4	

(5) Include your plot here:

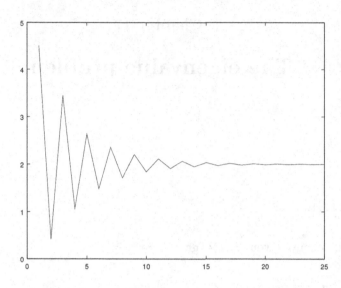

Which eigenvalue appears to be the limit? 2

Solution 17.2. :

(1) Include a copy of your **power_method.m** here:

```
function [r,x,rHistory,xHistory]=power_method( A, x ,maxNumSteps )
  % [r, x, rHistory, xHistory] = power_method( A, x ,maxNumSteps )
  % input variables:
  % A is the matrix
  % x is the starting vector
  % maxNumSteps is the maximum number of steps to take
  % output variables:
  % r is the final Rayleigh quotient (approximate eigenvalue)
  % x is the final value of x (approximate eigenvector)
  % rHistory(k) is the value of r on step k
  % xHistory(:,k) is the value of x on step k

  % M. Sussman

  x=x/norm(x);

  for k=1:maxNumSteps

    x=A*x/norm(A*x);
```

```
    r= x'*A*x;   %Rayleigh quotient

    rHistory(k)=r;      % save Rayleigh quotient on each step
    xHistory(:,k)=x;  % save one eigenvector on each step
  end
end
```

(2) Fill in the table

Power method iterations		
Step	Rayleigh q.	x(1)
0	-4	0.0
1	0.20455	0.12309
2	2.19403	0.51832
3	0.84889	0.27964
4	2.60494	0.53477
10	1.96240	0.46719
15	2.15332	0.44244
20	2.00887	0.44835
25	1.99789	0.44694

(3) Include your two plots here:

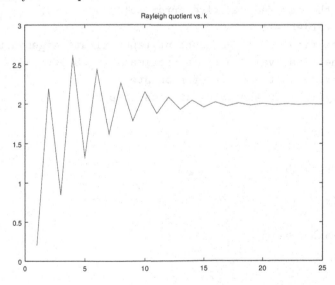

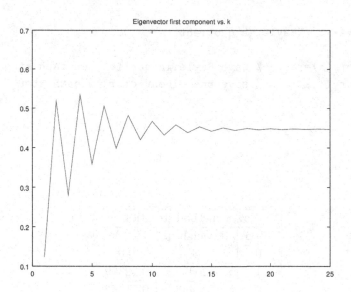

Eigenvector first component vs. k

Solution 17.3.

(1) Include a copy of your power_method1.m here:

```
function  [r, x, rHistory, xHistory] = power_method1 ( A, x , tol)
  % [r, x, rHistory, xHistory] = power_method ( A, x ,maxNumSteps )
  % input variables:
  % A is the matrix
  % x is the starting vector
  % tol is the desired relative error tolerance
  % output variables:
  % r is the final Rayleigh quotient (approximate eigenvalue)
  % x is the final value of x (approximate eigenvector)
  % rHistory(k) is the value of r on step k
  % xHistory(:,k) is the value of x on step k

  % M. Sussman

  maxNumSteps=10000;

  x=x/norm(x);

  for k=1:maxNumSteps

    x=A*x/norm(A*x);
```

```
    r= x'*A*x;   %Rayleigh quotient

    rHistory(k)=r;      % save Rayleigh quotient on each step
    xHistory(:,k)=x;    % save one eigenvector on each step

    if k>1
      if abs(r-rHistory(k-1)) <= tol*abs(r) & ...
            norm(x-xHistory(:,k-1)) <= tol
        return
      end
    end
  end
  disp('power_method1: Failed!')
end
```

(2) Fill in the following table.

Matrix	Eigenvalue	x(1)	no. iterations
1	2	0.44721	69
2	4.7321	0.21132	39
3	6.2749	0.55100	18
4	3.8019	0.23192	110
5			Failed
6	22.392	0.28868	16
7	3.9962	-0.012191	4653

(3) Include your plots for matrix 3 here:

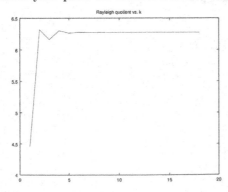

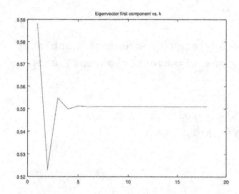

Include your plots for matrix 5 here:

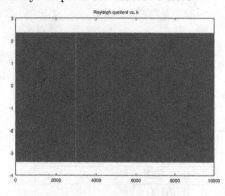

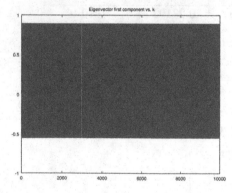

Plots for matrix 5, final 100 steps (not required):

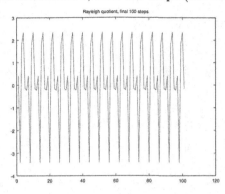

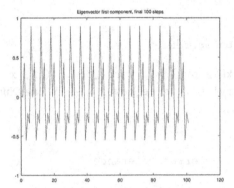

Include your plots for matrix 7 here:

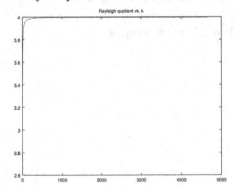

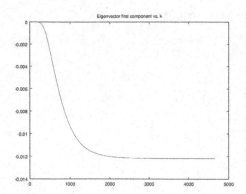

Solution 17.4.

(1) Include a copy of your inverse_power.m here:

```
function  [r, x, rHistory, xHistory] = inverse_power ( A, x , tol)
   % [r, x, rHistory, xHistory] = inverse_power ( A, x ,maxNumSteps )
   % input variables:
   % A is the matrix
   % x is the starting vector
   % tol is the desired relative error tolerance
   % output variables:
   % r is the final Rayleigh quotient (approximate eigenvalue)
   % x is the final value of x (approximate eigenvector)
   % rHistory(k) is the value of r on step k
   % xHistory(:,k) is the value of x on step k

   % M. Sussman

   maxNumSteps=10000;

   x=x/norm(x);

   for k=1:maxNumSteps

     x=A\x;
     x=x/norm(x);

     r= x'*A*x;   %Rayleigh quotient

     rHistory(k)=r;   % save Rayleigh quotient on each step
     xHistory(:,k)=x;  % save one eigenvector on each step
```

```
    if k>1
      if abs(r-rHistory(k-1)) <= tol*abs(r) & ...
             norm(x-xHistory(:,k-1)) <= tol
        return
      end
    end
  end
  disp('inverse_power: Failed!')
end
```

(2) What are **r**, **x** and the number of steps?
number of steps is 23, **r=1.2679**, **x=[0.78868; -0.57735; 0.21132]**.

(3) Is **r** close to **1/rp**? (<u>yes</u>/no)
Are **xHistory** and **xHp** close? (<u>yes</u>/no)
Are **rHistory** and **rHp** close at the end? (<u>yes</u>/no)

Solution 17.5.

(1) Nothing required
(2) Include your two plots here.

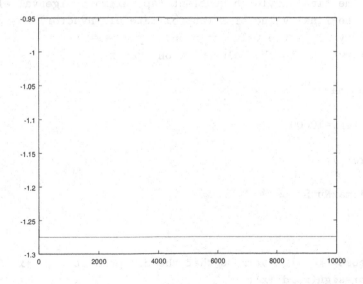

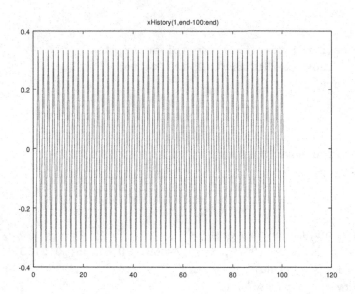

xHistory(1,end-100:end)

(3) Include your revised `inverse_power.m` here:

```
function  [r, x, rHistory, xHistory] = inverse_power ( A, x , tol)
  % [r, x, rHistory, xHistory] = inverse_power ( A, x ,maxNumSteps )
  % input variables:
  % A is the matrix
  % x is the starting vector
  % tol is the desired relative error tolerance
  % output variables:
  % r is the final Rayleigh quotient (approximate eigenvalue)
  % x is the final value of x (approximate eigenvector)
  % rHistory(k) is the value of r on step k
  % xHistory(:,k) is the value of x on step k

  % M. Sussman

  maxNumSteps=10000;

  x=x/norm(x);

  for k=1:maxNumSteps
    xold=x;

    x=A\x;
    % choose the sign of x so that the dot product xold'*x >0
    factor=sign(xold'*x);
```

```
   if factor == 0
     factor=1;
   end
   x=factor*x;


   x=x/norm(x);


   r= x'*A*x;   %Rayleigh quotient


   rHistory(k)=r;      % save Rayleigh quotient on each step
   xHistory(:,k)=x;   % save one eigenvector on each step


   if k>1
     if abs(r-rHistory(k-1)) <= tol*abs(r) & ...
            norm(x-xHistory(:,k-1)) <= tol
        return
     end
   end
  end
  disp('inverse_power: Failed!')
end
```

(4) Does `inverse_power` converge rapidly? (yes/no)

(5) Fill in the following table with the eigenvalue of smallest magnitude

matrix	Eigenvalue	x(1)	no. iterations
1	1.0000	-0.70711	50
2	1.2679	0.78868	23
3	-1.2749	-0.33415	40
4	0.19806	-0.23192	16
5			Failed
6	1.6077	0.28868	81
7	0.0037933	0.012191	15

Solution 17.6.

(1) Include a copy of your `power_several.m` here.

```
function  [r, V, numSteps] = power_several ( A, V , tol )
  % [r, V, numSteps] = power_several ( A, V , tol )
  % A is matrix
  % V is matrix whose columns are initial guesses for the
  %    several eigenvectors.
  % r is the vector of Rayleigh Quotients of eigenvectors
```

```
% V is matrix whose columns are eigenvectors

% M. Sussman

[V,unused]=qr(V,0);  % orthogonalization by qr
R=V'*A*V;            % Several Rayleigh quotients
r=diag(R);    % Only diagonal entries are of interest

maxNoSteps=10000;
for numSteps=1:maxNoSteps
  Vold=V;
  rold=r;
  V=A*V; %power iteration step
  [V,unused]=qr(V,0);  % orthogonalization by qr

  d=sign(diag(Vold'*V));       % signs of dot prods of cols
  D=diag(d);                   % diagonal matrix with +-1
  V=V*D;                       % Transform V

  R=V'*A*V;
  r=diag(R);

  if numSteps>1
    if norm(r-rold) <= tol*norm(r) & ...
       norm(V-Vold,'fro') <= tol*norm(V,'fro')
      return
    end
  end
end
error('power_several: Failed with too many steps')
end
```

(2) Do your results agree? (<u>yes</u>/no)
(3) What are the eigenvalues? 3.8019, 3.2470
   How many iterations are needed? 110
   Eigenvectors

```
V =
  -0.23192   -0.41791
  -0.41791   -0.52112
  -0.52112   -0.23192
  -0.52112    0.23192
  -0.41791    0.52112
  -0.23192    0.41791
```

(4) Is `AV=VR` approximately satisfied? (<u>yes</u>/no)

(5) How many iterations were needed? <u>106</u>

What are the six eigenvalues? 3.80194 3.24698 2.44504 1.55496 0.75302 0.19806

Solution 17.7.

(1) Include a copy of your `inverse_power.m` here:

```
function  [r, x, numSteps] = shifted_inverse ( A, x , shift, tol)
  % [r, x, numSteps] = shifted_inverse ( A, x , shift, tol)
  % shifted inverse power method for eigenvalues
  % input variables:
  % A is the matrix
  % x is the starting vector
  % shift is the amount of the shift
  % tol is the desired relative error tolerance
  % output variables:
  % r is the final Rayleigh quotient (approximate eigenvalue)
  % x is the final value of x (approximate eigenvector)
  % numSteps is number of steps taken

  % M. Sussman

  maxNumSteps=10000;

  x=x/norm(x);
  r=0;

  for numSteps=1:maxNumSteps
    rold=r;
    xold=x;

    x=(A-shift*eye(size(A)))\x;
    % choose the sign of x so that the dot product xold'*x >0
    factor=sign(xold'*x);
    if factor == 0
      factor=1;
    end
    x=factor*x;

    x=x/norm(x);

    r= x'*A*x;   %Rayleigh quotient
```

```
        if numSteps>1
           if abs(r-rold) <= tol*abs(r) & ...
                 norm(x-xold) <= tol
              return
           end
        end
     end
     disp('shfited_inverse: Failed!')
   end
```

(2) Are your results identical and take the same number of steps? (yes/no)
(3) How many iterations? 11
    Did you get the same eigenvalue and eigenvector? (yes/no)
(4) How many iterations? 4
    Does you result agree? (yes/no)
(5) How many iterations? 21
    What is the eigenvalue? 3
(6) What are the three eigenvalues and eigenvectors?

```
   r =   1.2679
   x =
      0.78868
     -0.57735
      0.21132

   r =  3
   x =
     -0.57735
     -0.57735
      0.57735

   r =   4.7321
   x =
      0.21132
      0.57735
      0.78868
```

Do they agree with the results of `eig`? (yes/no)

Solution 17.8.

(1) Include a copy of your `qr_method.m` here:

```
function [e,numSteps]=qr_method(A,tol)
   % [e,numSteps]=qr_method(A,tol) uses the qr method to find all
```

```
% eigenvalues of a matrix
% A is matrix
% tol is convergence tolerance
% e is vector of all eigenvalues

% M. Sussman

maxNumSteps=10000;
for numSteps=1:maxNumSteps
  [Q,R]=qr(A); % factorization
  Aold=A;
  A=R*Q;
  if norm(tril(A)-tril(Aold),'fro') <= tol*norm(tril(A),'fro')
    e=diag(A);
    return;
  end
end
error('qr_method: convergence failure')
end
```

(2) Did you get the same values? (yes/no)
(3) Try out the QR iteration for the **eigen_test** matrices except number 5. Use a
    relative tolerance of 1.0e-8. Report the results in the following table.

Matrix	largest eigenvalue	smallest eigenvalue	number of steps
1	2.0000	-1.5000	70
2	4.7321	1.2679	39
3	6.2749	-2.0000	39
4	3.80194	0.19806	95
(5)	skip	skip	skip
6	22.3923	1.6077	72
7	3.9962	0.0037933	2083

(4) For matrix 7, what is the difference between your computed largest eigenvalue
    and the one from the MATLAB **eig** function? 3.1835e-07
    For the smallest eigenvalue? 3.7557e-16
    How does the size of these errors compare with the tolerance of 1.0e-8?
    Largest eigenvalue is a little larger than tolerance, smallest is much smaller.
(5) What happens to the iterates?
    They seem to form a cycle.
(6) Convergence in 53 iterations.

```
e-(1+i)=
   1.00000 + 1.73205i
   1.00000 - 1.73205i
   2.00000
  -2.00000
```

Solution 17.9.

(1) Include a copy of your `qr_convergence.m` here:

```
function [e,numSteps]=qr_convergence(A,tol)
  % [e,numSteps]=qr_convergence(A,tol) uses the qr method to
  %    find all eigenvalues of a matrix
  % A is matrix
  % tol is convergence tolerance
  % e is vector of all eigenvalues

  % M. Sussman

  maxNumSteps=10000;
  for numSteps=1:maxNumSteps
    [Q,R]=qr(A); % factorization
    Aold=A;
    eold=diag(Aold);
    A=R*Q;
    e=diag(A);
    lambda=sort(abs(e));
    rho=max(lambda(1:end-1)./lambda(2:end));
    if norm(e-eold) <= tol*norm(e)*(1-rho);
       return;
    end
  end
  error('qr_convergence: convergence failure')
end
```

(2) Do the three eigenvalues approximately agree? (<u>yes</u>/no)
(3) What is the number of iterations? 50
    What are the eigenvalues? -4.1496, 3.3243, -2.1747
(4) How close are your eigenvalues? 3.0618e-09
    Is the relative error near the tolerance? (<u>yes</u>/no)
(5) Does the relative error decrease by a factor of ten? (<u>yes</u>/no)
(6) How many iterations? 2374
(7) What is the relative error? 4.9688e-09

(8) How many iterations for qr_method? 2083
Which convergence method yields more accurate results? qr_convergence
Which method would you use? qr_convergence
(9) Explain how you know it failed and what went wrong.
All iterates are the same, $\rho = 1$.

Solution 17.10.

(1) Include your copy of myroots.m here:

```
function r=myroots(a)
  % r=myroots(A) computes roots of polynomial with
  % coefficients a

  % M. Sussman

  N=length(a)-1;
  a=reshape(a,N+1,1); %force a to be acolumn vector

  A=diag(ones(N-1,1),-1); %put ones in lower diagonal
  A(:,end)=-a(N+1:-1:2);

  r=eig(A);
end
```

(2) Explain your tests here:
Part of the author's intention here is to assess the quality and comprehensiveness
of the tests. One example of a good test might be:

```
>> a=poly([1,-1,1+i,1-i,sqrt(2),1000])
a =
 Columns 1 through 6:
   1.0000 -1003.4142 3418.0420 -3827.8413 -590.6149 4831.2556
 Column 7:
  -2828.4271

>> r=myroots(a)
r =
  -1.0000e+00 + 0.0000e+00i
   1.0000e+00 + 1.0000e+00i
   1.0000e+00 - 1.0000e+00i
   1.0000e+00 + 0.0000e+00i
   1.4142e+00 + 0.0000e+00i
   1.0000e+03 + 0.0000e+00i
```

This includes two real roots of the same magnitude, a complex conjugate pair, a real root with the same magnitude as the complex roots, and a real root of comparatively larger magnitude.

Chapter 18

Singular value decomposition

Solution 18.1.
Include a copy of your **exer1.m** here:

```
% Chapter 18, exercise 1
% M. Sussman

% (1)
N=20;
d1=rand(N,1);
d2=rand(N,1);
d3=rand(N,1);
d4=4*d1-3*d2+2*d3-1;

% (2)
EPSILON=1.e-5;
d1=d1.*(1+EPSILON*rand(N,1));
d2=d2.*(1+EPSILON*rand(N,1));
d3=d3.*(1+EPSILON*rand(N,1));
d4=d4.*(1+EPSILON*rand(N,1));

% (3)
A=[d1,d2,d3,d4];

% (4)
[U S V] = svd(A);
fprintf('The four non-zero values of S= %e %e %e %e\n',diag(S));

% (5)
check=norm(A-U*S*V','fro')

% (6)
```

```
Splus=zeros(size(A'));
Splus(1:4,1:4)=diag(1./diag(S));

% (7)
x=V*Splus*U'*ones(N,1)
```

What is your solution?

```
x = [ 3.99998; -2.99999; 1.99999; -1.00000]
```

Is it close to the known solution? (yes/no)

Solution 18.2.

(1) Include a copy of your **exer2.m** here:

```
% Chapter 18, exercise 2
% M. Sussman

% (1)
N=20;
d1=rand(N,1);
d2=rand(N,1);
d3=d1+d2;
d4=4*d1-3*d2+2*d3-1;

% (2)
EPSILON=1.e-5;
d1=d1.*(1+EPSILON*rand(N,1));
d2=d2.*(1+EPSILON*rand(N,1));
d3=d3.*(1+EPSILON*rand(N,1));
d4=d4.*(1+EPSILON*rand(N,1));

% (3)
A=[d1,d2,d3,d4];

% (4)
[U S V] = svd(A);
fprintf('The four non-zero values of S= %e %e %e %e\n',diag(S));

% (5)
check=norm(A-U*S*V','fro')
```

```
% (6)
Splus=zeros(size(S'));
S(4,4)=0;
for k=1:3
  Splus(k,k)=1/S(k,k);
end

% (7)
x=V*Splus*U'*ones(N,1)

% fourth component of V(:,4)
% to find multiple of V(:,4) to add to x:
multiple=sum(([4;-3;2]-x(1:3))./V(1:3,4))/3
% check
x+multiple*V(:,4)
```

What solution did you find?
[4.33328; -2.66666; 1.66667; -0.99999]

(2) What multiple of V(:,4) can be added to your solution to yield [4;-3;2;-1]?
-0.57732

Solution 18.3.

(1) Nothing required for (1)–(4)
(5) Include your image plot here:

Grayscale NASA photo

(6) Include your singular value plot here:

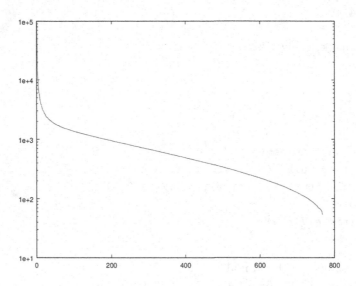

(7) Nothing required.

(8) Include the **nasa25** plot here:

Solution 18.4.

(1) Include a copy of your **jacobi_svd.m** here:

```
function [U,S,V]=jacobi_svd(A)
  % [U S V]=jacobi_svd(A)
  % A is original square matrix
  % Singular values come back in S (diag matrix)
  % orig matrix = U*S*V'
  %
  % One-sided Jacobi algorithm for SVD
  % lawn15: Demmel, Veselic, 1989,
  % Algorithm 4.1, p. 32

  % M. Sussman
  TOL=1.e-8;
  MAX_STEPS=40;

  n=size(A,1);
  U=A;
  V=eye(n);
  for steps=1:MAX_STEPS
    converge=0;
    for j=2:n
      for k=1:j-1
        % compute [alpha gamma;gamma beta]=(k,j) submatrix of U'*U
        alpha= sum( U(:,k).^2 );
        beta=  sum( U(:,j).^2 );
        gamma= sum( U(:,k).*U(:,j) );
        converge=max(converge,abs(gamma)/sqrt(alpha*beta));

        % compute Jacobi rotation that diagonalizes
        %    [alpha gamma;gamma beta]
        if gamma ~= 0
          zeta=(beta-alpha)/(2*gamma);
          t=sign(zeta)/(abs(zeta)+sqrt(1+zeta^2));
        else
          % if gamma=0, then zeta=infinity and t=0
          t=0;
        end
        c=1/sqrt(1+t^2);
        s=c*t;

        % update columns k and j of U
        T=U(:,k);
        U(:,k)=c*T-s*U(:,j);
```

```
            U(:,j)=s*T+c*U(:,j);

            % update matrix V of right singular vectors

            T=V(:,k);
            V(:,k)=c*T-s*V(:,j);
            V(:,j)=s*T+c*V(:,j);

        end
      end
      if converge < TOL
        break;
      end
    end
    if steps >= MAX_STEPS
      error('jacobi_svd failed to converge!');
    end

    % the singular values are the norms of the columns of U
    % the left singular vectors are the normalized columns of U
    for j=1:n
      singvals(j)=norm(U(:,j));
      U(:,j)=U(:,j)/singvals(j);
    end
    S=diag(singvals);
  end
```

(2) What are the matrices produced by your `jacobi_svd`?

```
U1 =
   0.60000   -0.80000
   0.80000    0.60000


V1 =
   0.70711   -0.70711
   0.70711    0.70711


S1 =
   5    0
   0    4
```

How close is the product of your computed U*S*V' to A?
`norm(A-U1*S1*V1','fro')/norm(A,'fro') = 7.7541e-17`

(3) What are your values for the matrices U1, S1, and V1?

U1 =
```
  -0.9475806816931032    0.2124354640016070   -0.2386671852527192
   0.3162370945482255    0.5168004524249504   -0.7955572841757300
   0.0456612714861348    0.8293301308934751    0.5568900989230110
```

V1 =
```
   0.838248076909180     0.513405188225107    -0.183726084869850
  -0.510062175291238     0.619109435357744    -0.597109775827013
  -0.192812704185257     0.594237847797124     0.780835860696863
```

S1 =
```
   1.13717372900606                    0                    0
                  0    16.75430798063765                    0
                  0                    0     1.73205080756888
```

How do they agree with ones computed by the MATLAB svd function? Columns
are permuted, some columns are negatives of others. Otherwise same.

Solution 18.5.

(1) Include a copy of your bidiag_reduction.m here:

```
function [U,B,V]=bidiag_reduction(A)
  % [U B V]=bidiag_reduction(A)
  % Algorithm 5.4.2 in Golub & Van Loan, Matrix Computations
  % Third Edition, Johns Hopkins University Press 1996.
  % Finds an upper bidiagonal matrix B so that A=U*B*V'
  % with U,V orthogonal.  A is an m x n matrix

  % M. Sussman

  [m,n]=size(A);
  B=A;
  U=eye(m);
  V=eye(n);
  for k=1:n-1
    % eliminate non-zeros below the diagonal
    % Keep the product U*B unchanged
    H=householder(B(:,k),k);
    B=H*B;
    U=U*H';

    % eliminate non-zeros to the right of the
```

```
        % superdiagonal by working with the transpose
        % Keep the product B*V' unchanged
        H=householder(B(k,:)',k+1);
        H=H';
        B=B*H;
        V=V*H;
    end
end
```

(2) Nothing required
(3) Include the matrices U, B, V and describe your tests to show they are correct.

```
U =
  -0.447214  -0.442554   0.604171  -0.462884  -0.157675
  -0.447214  -0.358347   0.018894   0.614928   0.541381
  -0.447214  -0.168645  -0.475469   0.213553  -0.707024
  -0.447214   0.187046  -0.519700  -0.566964   0.416506
  -0.447214   0.782500   0.372104   0.201367  -0.093188


B =
  -2.2361e+00   6.6672e+01  -9.9034e-15   8.5775e-15   7.4526e-15
  -5.0836e-18  -6.3646e+01   6.2904e+00   2.6504e-16   8.6391e-16
  -6.5679e-17  -1.1878e-15   2.8124e+00   6.9305e-01  -5.5511e-17
   5.0104e-17  -4.7094e-16   1.1433e-16   2.2060e-01  -5.3148e-02
   9.1272e-18   1.2539e-15  -2.0823e-16  -6.9389e-17   1.1326e-02


V =
   1.00000   0.00000   0.00000   0.00000   0.00000
   0.00000  -0.10061  -0.54181   0.75183   0.36203
   0.00000  -0.23477  -0.61419  -0.11553  -0.74452
   0.00000  -0.46953  -0.36848  -0.59014   0.54360
   0.00000  -0.84516   0.43982   0.27044  -0.13829
norm(U*B*V'-A,'fro')/norm(A,'fro') =    6.8987e-16
norm(U*U'-eye(5),'fro') =    8.4175e-16
norm(U'*U-eye(5),'fro') =    8.0307e-16
norm(V'*V-eye(5),'fro') =    7.1319e-16
norm(V*V'-eye(5),'fro') =    8.0473e-16
norm(B-diag(diag(B))-diag(diag(B,1),1),'fro') =    1.5208e-14
```

(4) Describe your tests to show your results are correct.

```
A=rand(100,100);
[U B V]=bidiag_reduction(A);
norm(U*B*V'-A,'fro')/norm(A,'fro') =    1.5703e-15
```

```
norm(U*U'-eye(100),'fro')  =    2.3235e-14
norm(U'*U-eye(100),'fro')  =    2.3266e-14
norm(V'*V-eye(100),'fro')  =    2.3562e-14
norm(V*V'-eye(100),'fro')  =    2.3301e-14
norm(B-diag(diag(B))-diag(diag(B,1),1),'fro')  =    2.6117e-14
```

Solution 18.6.

(1) Include a copy of your givens_rot.m here:

```
function [c,s,r]=givens_rot(f,g)
   % [c,s,r]=givens_rot(f,g) computes Givens rotation
   % [f;g] is a column vector
   % [c,s;-s,c]*[f;g]=[r;0] on exit

   % M. Sussman

   if f==0
      c=0;
      s=1;
      r=g;
   elseif abs(f) > abs(g)
      t=g/f;
      t1=sqrt(1+t^2);
      c=1/t1;
      s=t*c;
      r=f*t1;
   else
      t=f/g;
      t1=sqrt(1+t^2);
      s=1/t1;
      c=t*s;
      r=g*t1;
   end
   if norm([c,s;-s,c]*[f;g]-[r;0]) > 1.e-14*norm([f;g])
      error('givens_rot')
   end
end
```

(2) What values did you get? c=1
    s=0, r=1.

(3) What values did you get? c=0 s=1, r=2
    Does the matrix product yield [r;0]? (yes/no)

(4) What values did you get? c=0.44721 s=0.89443, r=2.2361.
Does the matrix product yield [r;0]? (yes/no)

(5) What values did you get? c=0.83205 s=-0.55470, r=-3.6056.
Does the matrix product yield [r;0]? (yes/no)

Solution 18.7.

(2) Include a copy of your msweep.m here:

```
function B=msweep(B)
  % B=msweep(B)
  % Demmel & Kahan zero-shift QR downward sweep
  % B starts as a bidiagonal matrix and is returned as
  % a bidiagonal matrix

  % M. Sussman

  n=size(B,1);
  for k=1:n-1
    [c s r]=givens_rot(B(k,k),B(k,k+1));

    % Construct matrix Q and multiply on the right by Q'.
    % Q annihilates both B(k-1,k+1) and B(k,k+1)
    % but makes B(k+1,k) non-zero.
    Q=eye(n);
    Q(k:k+1,k:k+1)=[c s;-s c];
    B=B*Q';
    % set almost-zero entries to true zero
    % display matrix and wait for a keypress
    B(find(abs(B)<1.e-13))=0;
    spy(B)
    disp('Plot completed. Strike a key to continue.')
    pause

    [c s r]=givens_rot(B(k,k),B(k+1,k));

    % Construct matrix Q and multiply on the left by Q.
    % Q annihilates B(k+1,k) but makes B(k,k+1) and
    % B(k,k+2) non-zero.
    Q=eye(n);
    Q(k:k+1,k:k+1)=[c s;-s c];
    B=Q*B;
    % set almost-zero entries to true zero
    % display matrix and wait for a keypress
```

```
B(find(abs(B)<1.e-13))=0;
spy(B)
disp('Plot completed. Strike a key to continue.')
pause


    end
end
```

(3) What is B(10,10)? 9.2260e-05

Include one of your plots here:

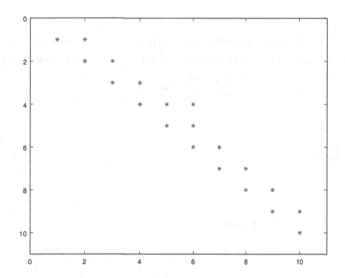

Solution 18.8.

(1) Include a copy of your vsweep.m here:

```
function [d,e]=vsweep(d,e)
  % [d,e]=vsweep(d,e)
  % Demmel & Kahan zero-shift QR downward sweep
  % d,e are diagonals of a bidiagonal matrix, e=upper diagonal

  % M. Sussman

  n=length(d);

  cold=1;
  c=1;
  for k=1:n-1
    [c,s,r]=givens_rot(c*d(k),e(k));
```

```
      if k ~= 1
        e(k-1)=r*sold;
      end
      [cold,sold,d(k)]=givens_rot(cold*r,d(k+1)*s);
    end
    h=c*d(n);
    e(n-1)=h*sold;
    d(n)=h*cold;
  end
```

(2) Do your results from **vsweep** agree with those from **msweep**? (yes/no)
(3) Do your results from **vsweep** agree with those from **msweep**? (yes/no)
    What is the time for **msweep**? 14.871
    What is the time for **vsweep**? 0.10277

Solution 18.9.

(1) Include your final plot here:

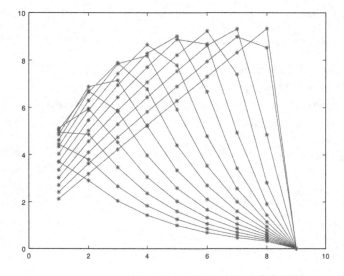

(2) Include your plot here:

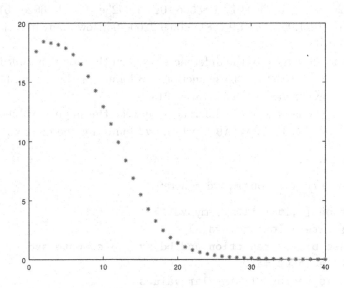

(3) Include your plot here:

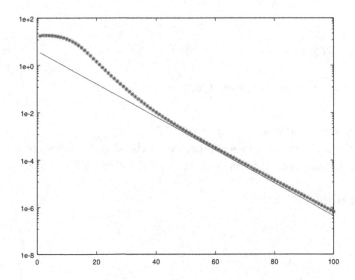

What is the value of r you found? 0.852

Solution 18.10.

(2) What are the singular values that `bd_svd` produces?
2.5422e+01 2.1675e+01 1.8836e+01 1.6525e+01 1.4602e+01 1.3001e+01
1.1671e+01 1.0574e+01 9.7881e+00 9.2260e-05 How many iterations did `bd_svd`
need? 276
What is the largest of the differences between the singular values `bd_svd` found
and those that the MATLAB function `svd` found for the matrix B? 1.2790e-13

(3) How many iterations does it take? 21368
What is the largest of the differences between the singular values `bd_svd` found
and those that the MATLAB function `svd` found for the matrix B? 1.0059e-13

Solution 18.11.

(1) Include a copy of your `mysvd.m` here:

```
function [d,iterations]=mysvd(A)
  % [d,iterations]=mysvd(A)
  % use bidiag_reduction and bd_svd to compute svd
  % A is input matrix
  % d is vector of singular values
  % iterations is number of iterations it took

  % M. Sussman

  [U B V]=bidiag_reduction(A);

  d=diag(B);
  e=diag(B,1);

  [d,iterations]=bd_svd(d,e);
end
```

(2) How many iterations are required? 257
How do the singular values compare with the singular values of A from the MAT-
LAB function `svd`? difference is at most 6.2528e-13 (occurs at largest singular
value)

(3) What is the rank of A? 7

Chapter 19

Iterative methods

Solution 19.1.

(1) Include your code for `meshplot.m` here:

```
N=10;
h=1/(N+1);
n=0;
% initialize x and y as column vectors
x=zeros(N^2,1);
y=zeros(N^2,1);
for j=1:N
  for k=1:N
    % xNode and yNode should be between 0 and 1
    xNode=j*h;
    yNode=k*h;
    n=n+1;
    x(n)=xNode;
    y(n)=yNode;
  end
end
plot(x,y,'o');
```

(2) Include your final code for `tests.m` here:

```
function tests(n)
  % function tests(n) prints index numbers of points near the
  % one numbered n

  % M. Sussman

  N=10;
  if mod(n,N)~=0
```

```
    nAbove=n+1
  else
    nAbove='none'
  end

  if mod(n,N)~=1
    nBelow=n-1
  else
    nBelow='none'
  end

  if n>N
    nLeft=n-N
  else
    nLeft='none'
  end

  if n<=N^2-N
    nRight=n+N
  else
    nRight='none'
  end
end %function
```

(a) With `tests(75)`, point above is <u>76</u> With `tests(80)`, point above is <u>none</u>
(b) With `tests(75)`, point below is <u>74</u> Pick some point with no point below <u>11</u>, `tests` gives <u>none</u>
(c) With `tests(75)`, point to left is <u>65</u> Pick some point with no point to the left <u>5</u>, `tests` gives <u>none</u>
(d) With `tests(75)`, point to right is <u>85</u> Pick some point with no point to the right <u>91</u>, `tests` gives <u>none</u>

Solution 19.2.

(1) Include `poissonmatrix.m` below;

```
function A=poissonmatrix(N)
  % A=poissonmatrix(N)
  % matrix generated by Equation (16.4)

  % M. Sussman

  h=1/(N+1);
  A=zeros(N^2,N^2);
```

```
    for n=1:N^2
      % center point value
      A(n,n)     =4/h^2;

      % "above" point value
      if mod(n,N)~=0
        nAbove=n+1;
        A(n,nAbove)=-1/h^2;
      end

      % "below" point value
      if mod(n,N)~=1
        nBelow= n-1;
        A(n,nBelow)=-1/h^2;
      end

      % "left" point value
      if n>N
        nLeft= n-N;
        A(n,nLeft)=-1/h^2;
      end

      % "right" point value
      if n<=N^2-N
        nRight= n+N;
        A(n,nRight)=-1/h^2;
      end
    end   %for
  end   %function
```

(2) What is `norm(A-A','fro')`? <u>0</u> Is it essentially zero? <u>yes</u>/no
(3) For a random vector v is `v'*A*v` positive? <u>yes</u>/no
(4) `det(A)` = <u>7.6579e+260</u>. Is it positive? <u>yes</u>/no
(5) Eigenvalue L= <u>48.219</u>, and eigenvector component E(10)= <u>-0.15232</u>, `norm(L*E-A*E,'fro')` = <u>3.2499e-13</u>. Is L*E approximately A*E? <u>yes</u>/no
(6) The smallest entry in the matrix `inv(A)` is <u>2.0830e-06</u>. It <u>is</u>/is not positive.
(7) `norm(gallery('poisson',N)/h^2-A,'fro')`= <u>0</u>. Is it essentially zero? <u>yes</u>/no

Solution 19.3.

(1) Include `anothermatrix.m` below:

```
function A=anothermatrix(N)
  % A=anothermatrix(N)
```

```
% matrix generated by Equation (16.4)

% M. Sussman

h=1/(N+1);
A=zeros(N^2,N^2);
for n=1:N^2
  % center point value
  A(n,n)      =8*n+2*N+2;

  % "above" point value
  if mod(n,N)~=0
    nAbove=n+1;
    A(n,nAbove)=-2*n-1;
  end

  % "below" point value
  if mod(n,N)~=1
    nBelow= n-1;
    A(n,nBelow)=-2*n+1;
  end

  % "left" point value
  if n>N
    nLeft= n-N;
    A(n,nLeft)=-2*n+N;
  end

  % "right" point value
  if n<=N^2-N
    nRight= n+N;
    A(n,nRight)=-2*n-N;
  end
 end %for
end %function
```

(2) For `A=anothermatrix(12)`, `norm(A-A','fro')` = <u>0</u>. Is it essentially zero?
 <u>yes</u>/no
(3) For a random vector v, `v'*A*v` = <u>1.0090e+04</u> Is it positive? <u>yes</u>/no
(4) Include your final plot here:

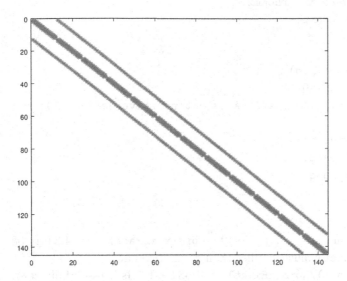

(5) The computed value for testing is 617237640.

Solution 19.4.

(1) Include your code for cgm.m here:

```
function x=cgm(A,b,x,m)
  % x=cgm(A,b,x,m)
  % A=spd matrix
  % b=right side
  % on input x=initial guess
  % on output x=soluiton
  % m=fixed number of iterations

  % M. Susssman

  r=b-A*x;
  rhoKm1=0;
  for k=1:m
    rhoKm2=rhoKm1;
    rhoKm1=dot(r,r);
    if rhoKm1 == 0
      return; % found solution
    end
    if k==1
      p=r;
    else
```

```
      beta=rhoKm1/rhoKm2;
      p=r+beta*p;
    end
    q=A*p;
    gamma=dot(p,q);
    if gamma <=0
      error('cgm: matrix A is not positive definite');
    end
    alpha=rhoKm1/gamma;
    x=x+alpha*p;
    r=r-alpha*q;
  end
end
```

(2) norm(y-xExact) =   7.1299e-12   norm(x-xExact) =   4.3431e-08.   Is
    norm(y-xExact) much smaller than norm(x-xExact)? yes/no
(3) norm(x-xExact)/norm(xExact) = 1.5841e-16. Is it essentially zero? yes/no
(4) For N=31, how close is the solution x after only one hundred steps? 4.2565e-11

Solution 19.5.

(1) Nothing required
(2) Include your plot here:

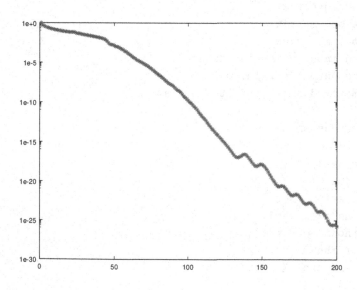

    How many iterations to get to 10^{-12}? 110
(3) Include your code for cg.m here:

```
function [x,k]=cg(A,b,x,tolerance)
  % x=cg(A,b,x,tolerance)
  % A=spd matrix
  % b=right side
  % on input x=initial guess
  % on output x=soluiton
  % k=number of iterations
  % tolerance=relative convergence tolerance

  % M. Susssman

  [rowsA,colsA]=size(A);
  if rowsA ~= colsA
    error('cg: A must be a square matrix')
  end
  [rows,cols]=size(x);
  if rows ~= rowsA | cols ~= 1
    error('cg: x must be a column vector compatible with A')
  end
  [rows,cols]=size(b);
  if rows ~= rowsA | cols ~= 1
    error('cg: b must be a column vector compatible with A')
  end

  r=b-A*x;
  rhoKm1=0;
  normBsquare=norm(b)^2;
  targetValue=tolerance^2*normBsquare;
  for k=1:rowsA
    rhoKm2=rhoKm1;
    rhoKm1=dot(r,r);
    if rhoKm1 == 0
      return; % found solution
    end
    if rhoKm1 < targetValue
      return;
    end
    if k==1
      p=r;
    else
      beta=rhoKm1/rhoKm2;
      p=r+beta*p;
```

```
      end
      q=A*p;
      gamma=dot(p,q);
      if gamma <=0
        error('cg: matrix A is not positive definite');
      end
      alpha=rhoKm1/gamma;
      x=x+alpha*p;
      r=r-alpha*q;
    end
  end
```

(4) (a) For tolerance of 1.0e-10, how many iterations does it take? <u>101</u>
    (b) What is the true error (`norm(x-xExact)/norm(xExact)`)? <u>4.2565e-11</u>
    (c) For tolerance of 1.0e-12, how many iterations does it take? <u>111</u>
    (d) What is the true error? <u>2.3980e-13</u>

(5) For `A=anothermatrix(N)` and tolerance of 1.e-10, How many iterations does it take? <u>162</u> What is the true error? <u>1.9182e-10</u>

Solution 19.6.

(1) Include your code for `anotherdiags.m` here:

```
function A=anotherdiags(N)
  % A=anotherdiags(N)
  % matrix generated by Equation (16.4)
  % using diagonal storage
  % main diag is 1st col, superdiag is 2nd col, far diag is 3d col

  % M. Sussman

  h=1/(N+1);
  A=zeros(N^2,3);
  for n=1:N^2
    % center point value: main diagonal
    A(n,1)    =8*n+2*N+2;

    % "above" point value: superdiagonal
    if mod(n,N)~=0
      A(n,2)=-2*n-1;
    end

    % "right" point value: "far" diagonal
    if n<=N^2-N
```

```
        A(n,3)=-2*n-N;
      end
    end
  end
```

(2)   `size(Adiags,2)-3 =` <u>0</u>
        `norm(Adiags(:,1)-diag(A)) =` <u>0</u>
        `norm(Adiags((1:N^2-1),2)-diag(A,1)) =` <u>0</u>
        `norm(Adiags((1:N^2-N),3)-diag(A,N)) =` <u>0</u>
        Are these four values essentially zero? <u>yes</u>/no

(3) Include your code for **poissondiags.m** here:

```
function A=poissondiags(N)
  % A=poissondiags(N)
  % matrix generated by Equation (16.4)
  % diagonal storage form
  % main diag is 1st col, superdiag is 2nd col, far diag is 3d col

  % M. Sussman

  h=1/(N+1);
  A=zeros(N^2,3);
  for n=1:N^2
    % center point value: main diagonal
    A(n,1)    =4/h^2;

    % "above" point value: superdiagonal
    if mod(n,N)~=0
      A(n,2)=-1/h^2;
    end

    % "right" point value: "far" diagonal
    if n<=N^2-N
      A(n,3)=-1/h^2;
    end
  end
end
```

        `size(Adiags,2)-3 =` <u>0</u>
        `norm(Adiags(:,1)-diag(A)) =` <u>0</u>
        `norm(Adiags((1:N^2-1),2)-diag(A,1)) =` <u>0</u>
        `norm(Adiags((1:N^2-N),3)-diag(A,N)) =` <u>0</u>
        Are these four values essentially zero? <u>yes</u>/no

Solution 19.7.

(1) Include your code for multdiags.m here:

```
function y=multdiags(A,x)
  % y=multdiags(A,x)
  % multiplication by diagonals

  % M. Sussman

  M=size(A,1);
  N=round(sqrt(M));
  if M~=N^2
    error('multdiags: matrix size is not a squared integer.')
  end
  if size(A,2) ~=3
    error('multdiags: matrix does not have 3 columns.')
  end
  if size(x,1) ~= M | size(x,2) ~=1
    error('multdiags: x is not a column vector consistent with A.')
  end

  % the diagonal product
  y=A(:,1).*x;
  % the superdiagonal product
  for k=1:M-1
    y(k)=y(k)+A(k,2)*x(k+1);
  end
  % the subdiagonal product
  for k=2:M
    y(k)=y(k)+A(k-1,2)*x(k-1);
  end
  % the far diagonal product
  for k=1:M-N
    y(k)=y(k)+A(k,3)*x(k+N);
  end
  % the far subdiagonal product
  for k=N+1:M
    y(k)=y(k)+A(k-N,3)*x(k-N);
  end
end
```

Warning for octave users! Octave can be very slow for this code. A much faster version, using elementary "vectorized" code is:

```
function y=multdiagsv(A,x)
  % y=multdiagsv(A,x)
  % diagonal storage, vectorized

  M=size(A,1);
  N=round(sqrt(M));
  if M~=N^2
    error('multdiags: matrix size is not a squared integer.')
  end
  if size(A,2) ~=3
    error('multdiags: matrix does not have 3 columns.')
  end
  if size(x,1) ~= M | size(x,2) ~=1
    error('multdiags: x is not a column vector consistent with A.')
  end

  % the diagonal product
  y=A(:,1).*x;
  % the superdiagonal product
  k=1:M-1;
  y(k)=y(k)+A(k,2).*x(k+1);

  % the subdiagonal product
  k=2:M;
  y(k)=y(k)+A(k-1,2).*x(k-1);

  % the far diagonal product
  k=1:M-N;
  y(k)=y(k)+A(k,3).*x(k+N);

  % the far subdiagonal product
  k=N+1:M;
  y(k)=y(k)+A(k-N,3).*x(k-N);

end
```

(2) What is the difference in results using multdiags to multiply anotherdiags by x=ones(9,1) and results of an ordinary multiplication by the matrix from anothermatrix.

```
norm(A*x-multdiags(Adiags,x)) = 0
```

(3) What is the difference in results using multdiags to multiply anotherdiags by x=(10:18)' and results of an ordinary multiplication by the matrix from

```
anothermatrix.

norm(A*x-multdiags(Adiags,x)) = 0
```

(4) Choose N=12, What is the difference in results using `multdiags` to multiply `anotherdiags` by x and results of an ordinary multiplication by the matrix from `anothermatrix`.

```
norm(A*x-multdiags(Adiags,x)) = 3.1672e-13
```

Solution 19.8.

(1) Include your code for `cgdiags.m` here:

```
function [x,k]=cg(A,b,x,tolerance)
  % x=cg(A,b,x,tolerance)
  % A=spd matrix
  % b=right side
  % on input x=initial guess
  % on output x=soluiton
  % k=number of iterations
  % tolerance=relative convergence tolerance

  % M. Susssman

  [rowsA,colsA]=size(A);
  if rowsA ~= colsA
    error('cg: A must be a square matrix')
  end
  [rows,cols]=size(x);
  if rows ~= rowsA | cols ~= 1
    error('cg: x must be a column vector compatible with A')
  end
  [rows,cols]=size(b);
  if rows ~= rowsA | cols ~= 1
    error('cg: b must be a column vector compatible with A')
  end

  r=b-A*x;
  rhoKm1=0;
  normBsquare=norm(b)^2;
  targetValue=tolerance^2*normBsquare;
  for k=1:colsA
    rhoKm2=rhoKm1;
    rhoKm1=dot(r,r);
```

```
      if rhoKm1 == 0
         return; % found solution
      end
      if rhoKm1 < targetValue
         return;
      end
      if k==1
         p=r;
      else
         beta=rhoKm1/rhoKm2;
         p=r+beta*p;
      end
      q=A*p;
      gamma=dot(p,q);
      if gamma <=0
         error('cgm: matrix A is not positive definite');
      end
      alpha=rhoKm1/gamma;
      x=x+alpha*p;
      r=r-alpha*q;
   end
end
```

(2) Test with N=3

Number of iterations using cg = <u>10</u>
Number of iterations using cgdiags = <u>10</u>
Relative norm of difference between the solutions = <u>1.1297e-16</u>

(3) Test with N=10, random right side

Number of iterations using cg = <u>65</u>
Number of iterations using cgdiags = <u>65</u>
Relative norm of difference between the solutions = <u>2.6017e-12</u>

(4) Contrived problem, N=500, known solution

How many iterations did it take? <u>672</u>
How much time did it take? <u>29 seconds</u>
What is the relative norm of the error? <u>2.5e-4</u>
How does the error compare with the tolerance? <u>larger</u>/smaller

Solution 19.9.

(1) How many entries are in Adiags? <u>750,000</u>
If one number takes eight bytes, how many bytes of central memory does Adiags take up? <u>6MB</u>

(2) The total size of `poissonmatrix(N)` is $(2.5\text{e}5)^\wedge 2 = 6.25\text{e}10$.

How many bytes of central memory would it take to store it? 500GB

How much would it cost to purchase enough memory to store that matrix? $1000

Does your estimate for memory cost alone exceed the cost of a home computer? yes/no

(3) C so that $T = CM^3$ holds is $C = 1\text{min}/(10^4)^3 = 10^{-12}\text{min}$

(4) How many days would it take to solve using ordinary storage? $C(500^2)^3 = 15,625$ min $= 10.85$ days

How does this compare with the time it took using storage by diagonals? Much longer than 30 seconds

Solution 19.10.

(1) What are $A(2,1)$ = 2 $A(3,2)$ = 2 and $A(5,1)$ = 0

(2) Indices and values of nonzero entries in column 4 are A(2,4)=5, A(3,4)=4, A(4,4)=3, A(5,4)=2, A(6,4)=1

Indices and values of nonzero entries in row 4 are A(4,2)=1, A(4,3=2), A(4,4)=3, A(4,5)=4, A(4,6)=5

(3) Sparse form of B is

(1,1)	1
(4,1)	5
(3,2)	4
(2,3)	3
(4,3)	6
(1,4)	2

Solution 19.11.

(1) Include your `precg.m` here:

```
function [x,k]=precg(U,A,b,x,tolerance)
  % x=precg(U,A,b,x,tolerance)
  % U is part of the preconditioner M=U'*U
  % A=spd matrix
  % b=right side
  % on input x=initial guess
  % on output x=soluiton
  % k=number of iterations
  % tolerance=relative convergence tolerance

  % M. Susssman

  [rowsA,colsA]=size(A);
```

```
if rowsA ~= colsA
  error('precg: A must be a square matrix')
end
[rowsU,colsU]=size(U);
if rowsU ~= colsU | rowsU ~= rowsA
  error('precg: U must be a square matrix, of same size as A')
end
[rows,cols]=size(x);
if rows ~= rowsA | cols ~= 1
  error('precg: x must be a column vector compatible with A')
end
[rows,cols]=size(b);
if rows ~= rowsA | cols ~= 1
  error('precg: b must be a column vector compatible with A')
end

r=b-A*x;
rhoKm1=0;
normBsquare=norm(b)^2;
targetValue=tolerance^2*normBsquare;
for k=1:colsA
  z=U\(U'\r);
  rhoKm2=rhoKm1;
  rhoKm1=dot(r,z);
  if rhoKm1 == 0
    return; % found solution
  end
  if dot(r,r) < targetValue
    return;
  end
  if k==1
    p=z;
  else
    beta=rhoKm1/rhoKm2;
    p=z+beta*p;
  end
  q=A*p;
  gamma=dot(p,q);
  if gamma <=0
    error('precg: matrix A is not positive definite');
  end
  alpha=rhoKm1/gamma;
```

```
    x=x+alpha*p;
    r=r-alpha*q;
  end
end
```

What MATLAB commands did you use to test your code?

```
A=poissonmatrix(10);
xExact=ones(100,1);
b=A*xExact;
x=zeros(100,1);
[y,k]=cg(A,b,x,1.e-10);
[yp,kp]=cg(A,b,x,1.e-10);
```

Both k-kp and norm(y-yp) are zero.

(2) Put your calculations here:

$$\mathbf{x}^0 = \mathbf{0}$$
$$M = A$$
$$\mathbf{r}^0 = \mathbf{b}$$
$$\mathbf{z} = A^{-1}b$$
$$\rho_0 = \mathbf{b} \cdot \mathbf{A}^{-1}\mathbf{b}$$
$$\mathbf{p} = A^{-1}\mathbf{b}$$
$$\mathbf{q} = A(A^{-1}\mathbf{b}) = \mathbf{b}$$
$$\gamma_1 = A^{-1}\mathbf{b} \cdot \mathbf{b}$$
$$\alpha_1 = \frac{\mathbf{b} \cdot \mathbf{A}^{-1}\mathbf{b}}{A^{-1}\mathbf{b} \cdot \mathbf{b}} = 1$$
$$\mathbf{x}^1 = \mathbf{p} = A^{-1}\mathbf{b}$$
$$\mathbf{r}^1 = \mathbf{b} - \mathbf{b} = \mathbf{0}, \text{so iteration will stop with k=2}$$

(3) Is the value of k correct? yes/no What MATLAB commands did you use to test your code? norm(y-xExact,'fro') gives 4.3170e-15

(4) How many iterations did conjugate gradient require? 222 What is the true error in the cg solution? 1.3692e-10 How many iterations did ICCG require? 36 What is the true error in the ICCG solution? 6.2242e-11 What MATLAB commands did you use to test your code?

```
N=50;
xExact=ones(N^2,1); % exact solution
A=anothermatrix(N);
b=A*xExact; % right side vector
tolerance=1.e-10;
U=chol(A); % Cholesky factor of A
x=zeros(N^2,1);
```

```
U(find(A==0))=0;
[y,n]=precg(U,A,b,x,tolerance);
precgIterations=n
precgError=norm(y-xExact)/norm(xExact)
[yy,nn]=cg(A,b,x,tolerance);
cgIterations=nn
cgError=norm(yy-xExact)/norm(xExact)
```

Instructions to Access the Supplementary Templates

To access the supplementary templates for this book, please follow the instructions below:

1. Register an account/login at https://www.worldscientific.com.

2. Go to: https://www.worldscientific.com/r/12340-supp to activate the access.

3. Access the templates from:

https://www.worldscientific.com/worldscibooks/10.1142/12340#t=suppl.

For subsequent access, simply log in with the same login details in order to access. For enquiries, please email: sales@wspc.com.sg.

Printed in the United States
by Baker & Taylor Publisher Services